Fundamentals of HVAC Control Systems

About the Authors

Ross D. Montgomery, P.E., CPMP, BEMP

An ASHRAE Vice-President, past DRC and DAL, Ross Montgomery has worked in ASHRAE activities for over 28 years, and has served on over 34 committees, councils and boards for ASHRAE. He has authored several Journal articles on controls and commissioning. Montgomery owns and operates QST-Inc, which specializes in commissioning/ controls, maintenance/operations, and energy audit/modeling. He also works as a certifying engineer for testing, adjusting, and balancing HVAC&R systems, as well as a primary investigator for energy/IAQ monitoring, testing, and research.

In addition to a BSME from the University of South Florida, Montgomery also has a Professional Engineers License, Mechanical/Electrical Contractors licenses, as well as numerous certifications, including IAQ, Commissioning, Energy Modeling, and Green Building Engineer. He is a past recipient of the ASHRAE Exceptional Service Award, Distinguished Service Award, John James International Award, an ASHRAE Technology Award, and the Lincoln Bouillon Membership Award.

Robert McDowall, P.Eng.

A professional engineer registered in Manitoba, Canada, Robert McDowall has been involved in building services for over forty years and has taught courses on Heating, Ventilation, and Air-Conditioning for thirty years. An active member of ASHRAE, Robert has served as a member of the Board, 1997-2000, as well as a member of the Standard 62 Committee. A subject matter expert in both HVAC systems and control systems, he is also the author of the online course and book Fundamentals of HVAC Systems and online course Fundamentals of HVAC Control Systems. Before undertaking these publications, Robert has worked as an engineering consultant for a variety of clients in the area of energy conservation and was responsible for the construction and maintenance of buildings for IBM and the University of Manitoba.

McDowall has both an MFM and MBA from the University of Manitoba, as well as an MA in Engineering from Cambridge University.

Fundamentals of HVAC Control Systems

Robert McDowell and Ross Montgomery are the authors of the
2009 edition of Fundamentals of HVAC Control Systems.

This book is a revision of an earlier edition authored by Steven T. Taylor in 1998.

180 Technology Parkway NW, Peachtree Corners, GA 30092

ISBN 978-1-933742-92-2

© 1998, 2011 ASHRAE
180 Technology Parkway NW
Peachtree Corners, GA 30092
www.ashrae.org

Library of Congress Cataloging-in-Publication Data

Montgomery, Ross.
 Fundamentals of HVAC control systems / Ross Montgomery, Robert McDowall. -- I-P ed.
    p. cm.
 Includes bibliographical references and index.
 Summary: "This text covers the need for HVAC controls, the basics of electricity, control valves and dampers, sensors and auxiliary devices, self- and system-powered controls, electric controls, pneumatic controls, analog electronic controls, diagrams and sequences, DDC hardware and software, DDC networks and control protocols, and digital control specification"–Provided by publisher.
  ISBN 978-1-933742-92-2 (hardcover)
  1. Heating--Control. 2. Ventilation--Control. 3. Air conditioning--Control.
 I. McDowall, Robert, P. Eng. II. Title.
  TH7466.5.M68 2011
  697--dc22
                        2010051135

Contents

Chapter 1

Introduction to HVAC Control Systems

Contents of Chapter 1

Study Objectives of Chapter 1

Chapter 1 introduces basic control concepts. It begins with a discussion of why controls are required in HVAC systems and a brief history of the development of control products. Next, we introduce the concept of a control loop, the basic building block of all control systems, and the various control strategies and algorithms used in control loops. After studying this chapter, you should understand:

Why controls are necessary in HVAC systems.

The difference between open and closed control loops.
How two-position, floating, and modulating control loops work.
Proportional control.
Integral and derivative control action in modulating control loops.
How to tune control loops.
The difference between direct acting and reverse acting.
Difference between normally open and normally closed.
How controlled devices may be sequenced using a single controller.

1.1 Why Do We Need Controls?

We need controls and control systems because, in our modern age of technology, they make our lives more convenient, comfortable, efficient, and effective. A control enables equipments to operate effectively and sometimes gives the ability to change their actions as time goes on and conditions or occupancies change. Controls can be devices used to monitor the inputs and regulate the output of systems and equipments. You use controls every day. For example, when you shower in the morning you sense the water temperature and manually modulate the hot and cold water valves to produce the desired temperature. When you drive to work, you monitor your speed using the speedometer and manually control the accelerator of your car to maintain the desired speed. When you get to your office, you sense a shortage of light so you manually switch on the overhead lighting.

These are all examples of closed-loop manual controls. The term *manual* means that you (a person, rather than a device) are acting as the controller; you are making the decisions about what control actions to take. The term *closed-loop* means that you have feedback from the actions you have taken. In these examples, the feedback comes from your senses of touch and sight: as you open the hot water valve in your shower, you can sense the temperature of the water increase; when you depress the accelerator, you can see that your speed is increasing by viewing the speedometer; and when you turn on the light, you can see that the brightness in the space has increased.

Your car may also be equipped with cruise control, to automatically maintain speed on a clear road, which is an example of an *automatic* control. An automatic control is simply a device that imitates the actions you would take during manual control. In this case, when you press the set-button on the cruise control panel, you are telling the controller the speed you desire, or the set point. The controller measures your speed and adjusts the position of the accelerator to attempt to maintain the car's speed at set point – the desired speed – just as you do when you manually control the speed.

You may notice that your cruise control system is able to maintain your car's speed at a given set point more precisely than you can manually. This is generally because you are not paying strict attention to controlling your speed; you must also steer, watch for traffic and perform all of the other functions required for safe driving. This is one reason why we use automatic controls: we do not have the time or desire, or perhaps the ability, to constantly monitor a process to maintain the desired result.

Controls of heating, ventilating and air-conditioning, and refrigerating (HVAC&R) systems are analogous in many ways to the controls we use to drive our cars. Just as we use speed as an indicator of safe driving, we generally use dry bulb temperature (the temperature that a common thermometer measures) as an indicator of comfortable thermal conditions. Just as speed is not the only factor that affects driving safety, temperature is not the only factor that affects our perception of thermal comfort. But like speed is the major factor in driving, temperature is the major factor in comfort and is readily measured and controlled. Your car's engine was designed to bring the car up to speed quickly, to drive it up a hill, or to carry a heavy load. But because we do not need this peak power output all of the time, we need a control device (the accelerator) that can regulate the engine's power output. The same can be said of HVAC systems. They are generally designed to handle peak cooling or heating loads that seldom, if ever, take place, so we must provide controls that can regulate the system's output to meet the actual cooling or heating load at a given time.

We use *automatic* controls for HVAC systems in place of manual controls, just as we might use cruise control to control the speed of our car. Automatic controls eliminate the need for constant human monitoring of a process, and therefore they reduce labor costs and provide more consistent, and often improved, performance.

The ultimate aim of every HVAC system and its controls is to provide a comfortable environment suitable for the process that is occurring in the facility. In most cases, the HVAC system's purpose is to provide thermal comfort for a building's occupants to create a more productive atmosphere (such as in an office) or to make a space more inviting to customers (such as in a retail store). The process may also be manufacturing with special requirements to ensure a quality product, or it may be a laboratory or hospital operating suite where, in addition to precise temperature and humidity control, the HVAC system must maintain room pressures at precise relationships relative to other rooms. With all of these systems, the HVAC system and its controls must regulate the movement of air and water, and the staging of heating, cooling, and humidification sources to regulate the environment.

Another capability that is expected of modern control systems is energy management. This means that while the control systems are providing the essential HVAC functions, they should do so in the most energy efficient manner possible.

Safety is another important function of automatic controls. Safety controls are those designed to protect the health and welfare of people in or around HVAC equipment, or in the spaces they serve, and to prevent inadvertent damage to the HVAC equipment itself. Examples of some safety control functions are: limits on high and low temperatures (overheating, freezing), limits on high and low pressures, freezestats, over current protection (e.g. fuses), and fire and smoke detection.

1.2 A Brief History of Controls

The first efforts at automatic control were to regulate space-heating systems. The bimetallic strip was the first device used; it controlled boiler output by opening and closing the boiler door, or a combustion air damper to control the rate of combustion. These devices were known as regulators. Other applications were to control steam radiators and steam heating coils. (Most steam radiators at that time were turned on and off by hand.)

Dr. Andrew Ure was probably the first person to call his regulator a thermostat and we still use this name 150 years later. These devices were soon used to control temperatures in incubators, railway cars, theaters, and restaurants.

Two other devices were developed to compete with the bimetallic strip. The first was a mercury thermometer column, having a contact low in the mercury and one or more contacts above the top of the column. Increasing temperature caused the mercury to rise and make contact with an upper electrode, thereby completing the circuit. This extremely accurate thermostat was nonadjustable.

The second device, a mercury switch uses a drop of mercury in a small, sealed glass tube with contacts at one or both ends. The horizontal glass tube is concave upwards, must be mounted level, and will make or break a circuit with a slight impulse from a bellows or bimetal sensor. This slight impulse is multiplied by the mass of the moving mercury. This device (discussed in Chapter 4) still is used to control countless HVAC systems.

Refrigeration systems used thermostats to cycle the motor driving the compressor, or to open and close valves to modulate capacity. The first refrigeration systems controlled the flow of refrigerant by hand. When smaller automatic equipment was developed, high side floats, low side floats, and constant pressure valves (automatic expansion valves) came into use.

These early control devices were generally electric; their function was to make or break an electric circuit that turned on a fan or pump, opened a valve or damper, etc. Some early controls (particularly burner controls on furnaces and boilers) were self-powered; meaning they drew their energy from the process itself rather than from an external source such as electricity. The need for inexpensive modulating controls (controls that could regulate output over a continuous range rather than cycling from full-on to full-off) lead to the development of pneumatic controls that use compressed air as the control power rather than electricity.

Pneumatic controls are inherently analog (modulating). With the invention of the electron tube, analog electronic controls were developed. These controls now use analog solid-state (semiconductor) devices to provide the desired control functions. Finally, with the emergence of powerful and inexpensive microprocessors, digital controls were developed. Digital controls (often called direct digital controls or DDC) use software programmed into circuits to effect control logic.

These five control system types—self-powered controls (described in Chapter 6), electric controls (Chapter 7), pneumatic controls (Chapter 8), analog electronic controls (Chapter 9), and digital controls (Chapter 10)—are the basis of modern control systems. Most control systems today use a combination of the five system types and are more accurately called hybrid control systems.

All of the various types of hardware used in temperature control systems (in the past, currently, and in the future) are based on the same fundamental principles of control. While the technology used to implement these principles may change, the fundamental concepts generally remain the same. These principles are the subject of the rest of this chapter.

(The historical information in this section is from the ASHRAE publication, *Heat and Cold: Mastering the Great Indoors.*)

1.3 Control Loops

The process of driving your car at a given speed is an example of a control loop. You use your speedometer to measure your car's speed. If you are below the desired speed, you press the accelerator and observe the response. If you continue below the desired speed, press the accelerator some more. As you approach the desired speed, you start to release the accelerator so that you do not overshoot it.

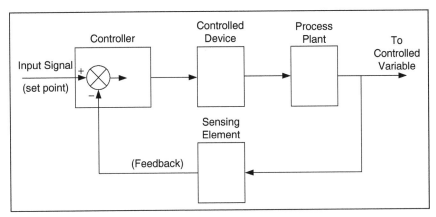

Figure 1-1   Diagram of Control Loop

In this example, you are acting as the controller making the control decisions whether to press or release the accelerator. The car's speed is the controlled variable and the speedometer is the sensor that measures the current value or control point of the controlled variable. The accelerator is the controlled device and your car's engine is the process plant.

Figure 1-1 shows this exchange of information schematically (see also *Table 1-1*).

It is called a control loop because information flows in a circle from the sensor (the speedometer) measuring the controlled variable (speed) to the controller (you), where the current value of the controlled variable (the control point) is compared to the desired value or set point. The controller then makes a control decision and passes that on to the controlled device (the accelerator) and to the process plant (the car's engine). This then has an effect on the current value or control point of the controlled variable, starting the process all over again. All control loops include these essential elements.

Figure 1-2 illustrates the components of a typical HVAC control loop.

Shown in *Figure 1-2* is an air-heating system utilizing a heating coil provided with steam, hot water, or some other heating source. Cold air is forced through the system using a fan and heated to some desired temperature to maintain its set point.

The intent of this control is to maintain a desired supply air temperature. As described in *Figure 1-2* and *Table 1-1*, the sensor measures the temperature of the supply air (the controlled variable) and transmits this information to the controller. In the controller, the measured temperature (the control point) is compared to the desired temperature (the set point). The difference between the set point and the control point is called the error. Using the error, the controller calculates an output signal and transmits that signal to the valve (the controlled device). As a result of the new signal, the valve changes position and changes the flow rate of the heating medium through the coil (the process plant). This, in turn, changes the temperature of the supply air. The sensor sends the new information to the controller and the cycle is repeated.

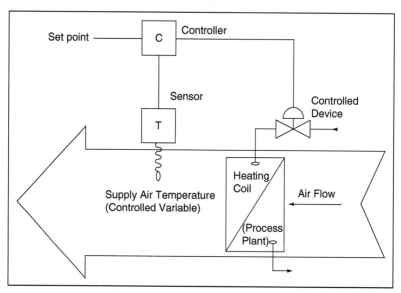

Figure 1-2   Simple Heating System

Table 1-1   Control Comparison for Automobile and Heating

Term	Automobile Example	Heating System Example	Definition
Controller	You	The device that provides a signal to the valve	The device that provides a signal to the controlled device in response to feedback from the sensor
Sensor	Speedometer	Supply air temperature sensor	The device that measures the current status of the controlled variable
Controlled device	The accelerator	The control valve	The device that changes the operation of the process plant in response to a control signal
Controlled variable	The car speed	The supply air temperature	The signal that the sensor senses
Process plant	The car engine	The heating coil	The device that produces the change in the controlled variable
Input signal (set point)	Desired speed	Supply air set point	This is the reference or desired input that is compared to the controlled variable

Both of these examples are called closed-loop or feedback-control systems because we are sensing the controlled variable and continuously feeding that information back to the controller. The controlled device and process plant have an effect on the controlled variable, which is sensed and fed to the controller for comparison to the set point and a subsequent response in the form of a change in controller output signal.

An open-loop control system does not have a direct link between the value of the controlled variable and the controller: there is no feedback. An example of an open-loop control would be if the sensor measured the outside air temperature and the controller was designed to actuate the control valve as a function of only the outdoor temperature. The variable (in this case, the supply air temperature or perhaps the temperature of the space the system served) is not transmitted to the controller, so the controller has no direct knowledge of the impact that valve modulation has on these temperatures modulating the valve.

Another way of defining an open-loop is to say that changes to the controlled device (the control valve) have no direct impact on the variable that is sensed by the controller (the outdoor air temperature in this case). With an open-loop control system, there is a presumed indirect connection between the end-result and the variable sensed by the controller.

If the exact relationship between the outdoor air temperature and the heating load was known, then this open-loop control could accurately maintain a constant space temperature. In practice, this is rarely the case, and therefore simple open-loop control seldom results in satisfactory performance. For this reason, almost all HVAC continuous-control systems use closed control loops.

Open-loop control, in the form of time-clocks, or occupancy sensors, are very common but they are on/off and not continuous controls. One form of open-loop control commonly used is called reset control. In reset control, an open-loop is used to provide a varying set point for a closed control loop. For example, an open-loop can be arranged to adjust the heating supply water temperature based on outside temperature, as shown in *Figure 1-3*. As the outside temperature falls, the open-loop output rises based on a predetermined schedule, shown in the table in *Figure 1-3*. This open-loop output provides the set point for the boiler.

The advantage of this reset control is that the capacity of the heating system increases as the load increases, greatly improving controllability. This use of one control loop to provide input to a second control loop is generally called cascading and other examples will be mentioned later in the course.

These examples describe the essential elements in any control loop: sensor, controller, controlled device, and process plant. Very few control systems are as simple as these examples, but every control system must include these

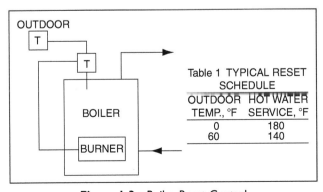

Table 1 TYPICAL RESET SCHEDULE	
OUTDOOR TEMP., °F	HOT WATER SERVICE, °F
0	180
60	140

Figure 1-3   Boiler Reset Control

essential elements. In Chapter 5, we will describe more complex systems, but all of them will be based on elementary control loops. Behind any apparent complexity, there must be an elementary system or systems.

Sometimes the sensor and controller are combined in one package. This sensor/controller combination is commonly called a stat, such as a thermostat, humidistat, or pressure stat. These devices still contain the individual control elements (the sensor and the controller); they have simply been mounted into a single enclosure.

Common controlled devices include control valves, which are used to control the flow of water or steam, and control dampers, which are used to control the flow of air. These devices and their proper selection are discussed in Chapter 3. Motor starters, relays, and variable speed drives are also common control devices; they are discussed in Chapter 2.

Common controlled variables include the temperature, humidity, pressure, and velocity of air in conditioned spaces and in ductwork, and the temperature, velocity, and pressure of water in hydronic heating and cooling systems. Sensors used to measure these variables come in a variety of types with varying degrees of accuracy. The accuracy of the measurement naturally affects the accuracy of control. Sensors are discussed in Chapter 4.

Typically, between the controller and the controlled device, there is an actuator attached to the controlled device through connectors called the linkage. Actuators are devices that convert the signal from the controller into a physical force that causes the controlled device (damper or valve) to move. Actuator characteristics vary by the type of control system used and are discussed in Chapters 6–10.

This section introduced some fundamental terms used in control systems. It is very important that these terms are well understood so that you understand the remaining sections of this chapter. The following is a summary of key terms:

Controlled variable: the property that is to be controlled, such as temperature, humidity, velocity, flow, and pressure.

Control point: the current condition or value of the controlled variable.

Set point: the desired condition or value of the controlled variable.

Sensor: the device that senses the condition or value of the controlled (or "sensed") variable.

Sensed variable: the property (temperature, pressure, humidity) that is being measured. Usually the same as the controlled variable in closed-loop control systems.

Controlled device: the device that is used to vary the output of the process plant, such as a valve, damper, or motor control.

Process plant: the apparatus or equipment used to change the value of the controlled variable, such as a heating or cooling coil or fan.

Controller: the device that compares the input from the sensor with the set point, determines a response for corrective action, and then sends this signal to the controlled device.

Control loop: the collection of sensor, controlled device, process plant, and controller.

Closed-loop: a control loop where the sensor is measuring the value of the controlled variable, providing feedback to the controller of the effect of its action.

Open-loop: a control loop where the sensor is measuring something other than the controlled variable. Changes to the controlled device and process plant have no direct impact on the controlled variable. There is an "assumed" relationship between the property that is measured and the actual variable that is being controlled.

1.4 Control Modes

The purpose of any closed-loop controller is to maintain the controlled variable at the desired set point. All controllers are designed to take action in the form of an output signal to the controlled device. The output signal is a function of the error signal, which is the difference between the control point and the set point. The type of action the controller takes is called the control mode or control logic, of which there are three basic types:

- two-position control
- floating control
- modulating control

Within each control-mode category, there are subcategories for each of the specific control algorithms (procedures, methods) used to generate the output signal from the error signal, and other enhancements used to improve accuracy.

The various control modes and subcategories are simply different ways of achieving the desired result: that the controlled variable be maintained at set point. It is often a difficult task because the dynamics of the HVAC system, the spaces it serves, and the controls themselves can be very complex. There are generally time lags between the action taken by the controller and the response sensed by the sensor.

For instance, in the heating system depicted in *Figure 1-2*, it takes time for the valve to move when given a signal from the controller; then it takes a little more time for the water to begin to flow. If the coil has not been used for a while, it will take time to warm its mass before it can begin warming the air. If the sensor was located in the space served by the system (as opposed to in the supply duct, as shown in *Figure 1-2*), there would be even more time delays as the air travels down the duct to the diffusers and begins to mix with and warm the air in the space. There will be a delay as the air warms the surfaces in the space (the space envelope, walls, and furnishings) before it warms the surfaces and the air near the sensor. There can also be a delay as the sensor itself takes time to warm up and reach a new steady-state condition, and there may be a delay in transferring that information back to the controller. Finally, the controller can take time to compare the sensor signal to the set point and calculate a response. The effect of all these delays is called the system time constant. If the time constant is short, the system will react quickly to a change in the controlled device or process plant; if the time constant is long, the system will be sluggish to changes in the controlling devices.

Another factor that affects the performance of a controller is the system gain. As shown in *Table 1-2*, the controller gain describes how much the controlled device will change for a given change in the controlled device and process plant.

Table 1-2   Controller Gain

Controller Setting	Controller Gain	Control Action
Open valve from 0 to 100% for a 1°F change in measured variable	Higher	Small change in measured variable creates a big change in output. 1°F causes controller to request 100% valve opening
Open valve from 0 to 100% for a 5°F change in measured variable	Lower	Large change in measured variable required to create a significant change in output. 1°F causes controller to request only 20% valve opening

The gain of the controller is only part of the issue. The capacity of the controlled device also influences the system gain. Thus, at full load, if full output just matches the requirement, then the controller gain is the same as the system gain. However, let us imagine that the required output is met with the controlled device only half open. In effect, the system capacity is twice what is required. The system gain is now the controller gain divided by the system overcapacity i.e., the gain is doubled. Thus, system gain is a function of controller gain and a function of how much capacity the system has relative to the load the system experiences. A system with a high gain means that a small change in the signal to the controlled device will cause a large change in the controlled variable. In the extreme case, the system is said to be oversized and good control is virtually impossible as even a small change in measured variable produces a huge change in output. The outdoor reset shown in *Figure 1-3* is an example of adjusting the system capacity to avoid the effect of overcapacity at low loads.

If all of these time delays and gains are fairly linear and consistent, the controller can generally be adjusted (tuned) to provide accurate control, but, there may be nonlinearities, such as hysteresis (delay, or uneven response) in the control valve. In this example, this is a delay caused by friction that binds the actuator or valve stem, thereby preventing a smooth, linear movement. There may also be changes to the system that change its gain, such as a varying hot water supply temperature to the heating coil, or a varying airflow rate through the coil. These complications can be handled with varying degrees of success and accuracy by the various control modes described in the following sections.

Two-position Control

The simplest and probably most common control mode is two-position control. It applies to systems that have only two states, such as On and Off for a fan or pump, or Open and Closed for a valve or damper. Small HVAC systems, such as the furnaces and air conditioners found in most residences, are examples of two-position systems. Systems that only have two states of operation are almost always controlled using two-position controls.

Figure 1-4 shows the action of a two-position control of a heating thermostat for the heating system.

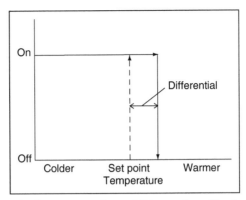

Figure 1-4   Two-position Control Diagram for a Heating System

Figure 1-5 shows what happens with this control when the controller responds to supply air that is colder than the desired set point at about 50% load.

The two positions in this case are the valve full-open (allowing full flow of hot water or steam through the coil) and the valve full-closed (no flow). Along the vertical axis is the value of the controlled (or "sensed") variable (the supply air temperature), while time is on the horizontal axis.

Looking again at *Figure 1-2* and starting at the left of *Figure 1-5*, because the air entering the coil is colder than the desired set point, the air temperature sensed by the supply temperature sensor begins to fall. Just as it falls below the set point (represented by the lower dashed horizontal line in the figure), the controller causes the valve to open. The heating medium flows to the coil, but the air temperature will not rise immediately because it takes time for the actuator to open the valve and for the coil mass to begin to warm before it starts to warm the air. So the air temperature continues to fall below the set point before it turns around and begins to rise.

The valve will stay open until the supply air temperature rises by the control differential above the set point. The control differential is a fixed difference in sensed value between the open and closed commands; it is represented in *Figure 1-5* by the difference between the set point and the higher dashed

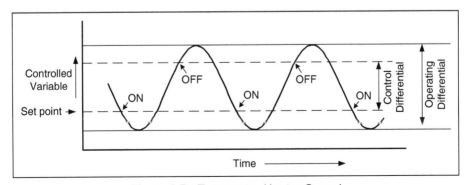

Figure 1-5   Two-position Heating Control

horizontal line. When the air temperature intersects this line, the controller closes the valve. Again, there is a time delay before the air temperature begins to fall because the heating medium in the coil and the coil itself must first cool off. Eventually, the air temperature falls until it reaches the set point again, and the cycle is repeated. In this example, the set point is shown to be the on-point (the point where the valve is opened) and the set point plus control differential is the off-point (where the valve is closed). It is also common to show the set point as being the midpoint between the on-point and off-point because this is the average condition of the controlled variable as the valve cycles opened and closed. In practice, however, the set point for a typical two-position thermostat (the point set by adjusting the set point knob) is usually either the on-point (as shown in *Figure 1-5*) or the off-point. It very seldom corresponds to the midpoint of the differential.

The overshoot and undershoot caused by the time delays result in the operating differential. This is depicted in *Figure 1-6* by the two solid horizontal lines marking the difference between the maximum and minimum temperature seen by the controlled variable. The operating differential always will be greater than the control differential.

It is not possible to remove the natural time-delays and thermal lag inherent in all real HVAC systems, but it is possible to reduce the difference between operating differential and control differential through the use of anticipation devices. One type of heat anticipator common in single-zone heating thermostats is a small resistance heater placed adjacent to the temperature sensor. This heater is energized when the heating is turned on and provides a false reading that causes the sensor to respond more rapidly. This causes the heater to be shut off before the actual space temperature rises above the control differential, reducing the over-shoot and thus reducing the operating differential. This is depicted in *Figure 1-6*.

The same device can be used to provide anticipation for cooling as well, except that in this case the heater is energized during the off-cycle. This causes

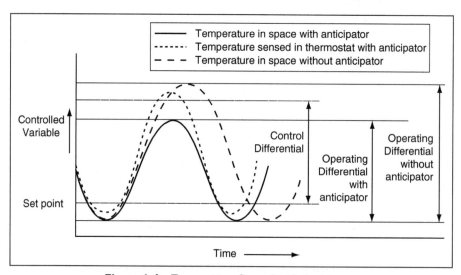

Figure 1-6   Two-position Control with Anticipator

the sensor signal to rise more quickly above set point, which then turns the cooling on more quickly, again reducing the operating differential.

Despite the use of anticipation devices, the very existence of a control differential results in temperature fluctuations. The air temperature is only at the desired condition (the set point) for a few moments. Control could be made more accurate by decreasing the control differential, but too small a differential will result in rapid cycling, called short-cycling. Short-cycling usually leads to inefficiencies in the heating or cooling system, and almost always shortens the life of equipment.

The ability of two-position control to function well (its ability to maintain the controlled variable near set point with a reasonable operating differential without short-cycling) is a function of the system gain, which is a function of the design of the HVAC system the control system. The capacity of the process plant (the heating or cooling system) must not greatly exceed the actual load experienced. If it does, then either the control differential must be increased, resulting in an unacceptably wide operating differential, or the system must be allowed to short-cycle. As with many control applications, the control system cannot compensate for a poor HVAC system design.

Where a wide variation in load is expected, if the system gain is too high due to process plant capacity that far exceeds the load at any given time, an HVAC system with continuously varying capacity or multiple capacity steps should be used. Systems with continuous capacity are controlled using either floating or modulating controls, as discussed in the next two sections. Systems with multiple capacity steps are controlled by step controls.

Step control is actually a series of two-position controls controlling the same controlled variable but at slightly different set points. Step control is used to control systems with multiple stages of capacity, such as multi-speed motors (high–low–off), multi-stage gas burners (high-fire, low-fire, off), or multi-stage refrigeration systems (multiple compressors, multi-speed compressors, or compressors with cylinder unloading). *Figure 1-7* shows how a step controller

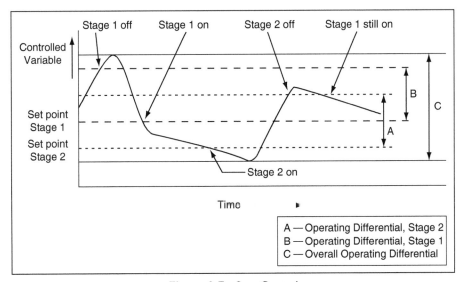

Figure 1-7   Step Control

might operate with a heating system that had two heating stages for a total of three operating positions (high-fire, low-fire, and off).

As the temperature falls below the Stage 1 set point, the first step of heating is turned on. If this were enough capacity to handle the load, the temperature would cycle around this set point and its control differential just as it would with two-position control. If, as depicted in *Figure 1-7*, the temperature were to continue to fall below the Stage 2 set point, the second heating stage would be turned on. Operation is then just like two-position control around this set point. Note that the control operating ranges overlap; the temperature at which the first stage comes on is lower than the temperature at which the second stage goes off. The set points and control differentials could have been set for nonoverlapping ranges where the first-stage on-point was equal to or above the second stage off-point. However, overlapping the ranges reduces the overall operating differential without using small operating differentials for each stage. In this way, fine control can be provided without short-cycling stages.

Floating Control

Floating control (also called "tri-state" control) is similar to two-position control, but the system it controls is not limited to two states. The system must have a modulating-type controlled device, typically a damper or valve driven by a bi-directional actuator (motor). The controller has three modes: drive open, idle (no movement), or drive closed.

Like two-position control, floating control has a set point and a control differential. Some floating controllers have, instead, two set point adjustments, an upper set point and a lower set point. The control differential is then the difference between the two set points.

When the supply air temperature falls below the lower line of the differential (in *Figure 1-8*), the controller starts to drive the control valve open, thereby increasing the flow of the heating medium through the coil. Because of the time delay in moving the valve and the normal thermal lags, the supply air temperature will continue to fall below the differential, but eventually it will start to rise. When it rises above the differential, the valve is no longer driven

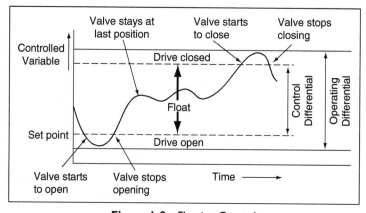

Figure 1-8   Floating Control

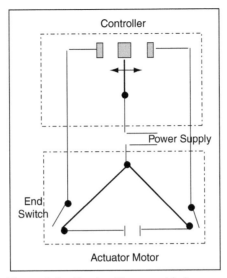

Figure 1-9   Floating Diagram with Actuator

open; it is left in the position it was in just when the temperature rose above the lower range of the differential.

While the temperature is within the differential range, the valve position will not move. If the temperature rises above the upper line of the differential, the controller will begin to drive the valve closed, thereby decreasing the flow of the heating medium through the coil. Again, the valve stops moving in whatever position it happens to be in once the temperature returns to within the differential. The air temperature will "float" within the differential range, which is how this control logic was given its name. Basically, as shown in *Figure 1-9*, a drive signal is applied to drive the actuator to either its open or closed position as the controller monitors its feedback.

Because the controlled device is variable rather than two-position, floating control can have a smaller control differential than two-position controllers without instability or excessive cycling. As with two-position control, floating control will have an operating differential that exceeds the control differential due to thermal lags and other time delays. In fact, the overshoot/undershoot can be even more severe because the valve timing (the time it takes to drive the valve from full-open to full-closed) must be relatively slow to prevent unstable control. (If the valve closes too quickly, the system will essentially behave as a two-position control.) For this reason, anticipation devices as described for two-position controls are especially desirable for floating controls.

Modulating Control

Can you imagine driving a car using two-position control? When your speed falls below the desired speed, you would fully depress the accelerator, and then release it at some differential above the desired speed. The results would be very jerky. To control speed precisely in a car, given the variation in power required from accelerating onto the freeway to coasting down a residential lane, a system of continuously varying capacity control is required: the accelerator. Many HVAC systems face similar widely varying loads and are also fitted with continuous or nearly continuous capacity capability. When controlling systems with this capability, you can improve the accuracy of control by using modulating control logic, the subject of this section.

Note that just because the system has a continuously variable output capability it does not limit you to using modulating controls. You could still drive your car with two-position control by alternately pressing and releasing the accelerator (although with not very satisfactory results). Early cruise control systems used floating control logic to control speed (as do some poor drivers, it seems). Conversely, systems that are inherently two-position or multi-position (such as staged refrigeration systems) are not limited to using two-position controllers. As we will see below, these systems can also benefit from modulating control techniques.

Modulating control is sometimes called analog control, drawing on the parallel between modulating/two-position and analog/digital. For many years, the term *proportional control* was used to mean modulating control because the controllers at that time were limited to proportional control logic (described below). Modern modulating controls use more sophisticated algorithms that go beyond simple proportional logic.

Remember that automatic controls simply imitate the human logic you would apply during manual control. To understand some of the control logic you use without consciously thinking about it, imagine you are trying to drive your car at a constant speed. While the road is flat or at a steady slope you can maintain a constant speed while pressing the accelerator to a fixed position. As you start to climb a hill, you begin to lose speed, so you begin to press the accelerator some more. If you slowed only a little below your desired speed, you may only press the accelerator a little more. The further you fall below the desired speed, the more you would press the accelerator.

After a while, you may find that you are no longer decelerating, but the speed you are maintaining, while constant, is still below that desired. The longer you stay below the desired speed, the more inclined you are to press the accelerator more to get back up to speed. However, let us say that, before you get up to speed, the terrain changes and you start to go down a hill. Your speed begins to increase toward the desired speed, but at a rate that you sense is too fast. To prevent from shooting past the speed you desire, you begin to back off of the accelerator even though you are still below the desired speed. As the terrain changes, you continuously go through many of these same thought processes over and again.

While you certainly do not think of them this way, the thought processes you use to drive your car can be approximated mathematically as:

$$V = V_0 + V_p + V_i + V_d$$
$$= V_0 + K_p e + K_i \int e\, dt + K_d \frac{de}{dt} \qquad\qquad (Equation\ 1\text{-}1)$$

In this expression, V is the output of the controller. Using our example, V is how much you press the accelerator.

The first term on the right-hand side of the equation, V_0, is called the offset adjustment. It is the amount you have to press the accelerator when you are driving on a flat road or a road with a steady slope, and to keep your car cruising steadily at the speed you desire.

The second term, V_p, is called the proportional term. It is proportional to the error e, which is the difference between actual speeds and the desired speed or set point. When you sense that the further you are from the desired speed, the more you should press the accelerator, you are using proportional control logic.

The third term, V_i, is called the integral term. It is proportional to the integral of the error over time. For those of you not familiar with calculus, the integral term is essentially a time-weighted average of the error; how much are you away from set point multiplied by how long you have been that way. In our example, when you sensed that you stayed below the desired speed for too long and thus pressed the accelerator more to increase speed, you were using integral control logic.

The last term V_d, is called the derivative term. It is proportional to the derivative of the error with respect to time. Again, if you are not familiar with calculus, the derivative term is essentially the rate of change of the error; how fast you are approaching or going away from set point. In our example, when you sensed that you were approaching the desired speed too quickly and thus started to back-off the accelerator, you were using derivative control logic.

These three terms can be seen graphically in *Figure 1-10*. The proportional term varies proportionally to the error: how far we are from set point. The integral term is proportional to the time-weighted average of the error, which is the area under the curve (the hatched area in the figure) and represents both how long and how far we have been away from set point. The derivative term is proportional to the slope of the error line; how quickly we are approaching or going away from set point. Now let us examine each of these terms to see how they affect control accuracy.

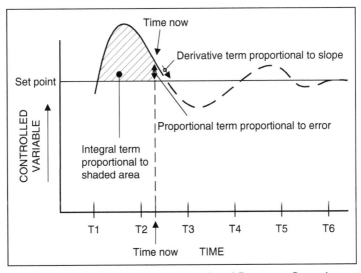

Figure 1-10   Proportional, Integral, and Derivative Control

First, if we remove the integral and derivative terms from *Equation 1-1*, we get:

$$V = V_0 + K_p e$$

(*Equation 1-2*)

This is the mathematical expression of proportional-only control logic. Proportional control is the simplest and most common modulating control logic. Virtually, all pneumatic thermostats, most pneumatic controllers, and most analog electronic controllers use it.

Figure 1-11 shows a typical proportional system response to start-up or change of set point. The system will respond by approaching the set point and then overshooting, due to the time delays and thermal lag mentioned under two-position control. Overshoot and undershoot will decrease over time until, under stable loads, the system levels out at some continuous value of the error (called offset or droop).

Continuous offset under steady-state conditions (constant load) is an inherent characteristic of proportional control. Proportional control will only keep the controlled variable exactly at set point under one specific load condition; at all others, there will be droop or offset.

Applying proportional logic and *Equation 1-2* to the heating coil depicted in *Figure 1-2*, you can see the following.

If we were at a steady load, then we would require a certain flow of the heating medium that would exactly match that load. To get that flow rate, we would have to open the valve to a certain position. That position is determined by the signal from the controller *V* in *Equation 1-2*.

If we assume that we were exactly at set point, then the second term in *Equation 1-2* would be zero because the error *e* would be zero, so the control signal would be equal to V_0 (our offset adjustment). For this specific load, we could adjust V_0 so that the desired flow rate is achieved and could maintain zero offsets. However, because V_0 is a constant, we can only maintain this condition at this precise load. If we were at a steady load that was higher, for example, we would need to open the valve more, requiring a larger signal *V*. To increase the value of *V*, the second term of *Equation 1-2* would have to be

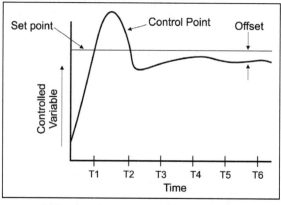

Figure 1-11   Proportional-only Control

nonzero, meaning our error would have to be nonzero. This error is the offset or droop.

The amount of the offset is a function of the constant K_p (the controller proportional gain). The larger the gain is, the smaller the offset. However, increasing the gain to minimize offset must be done with care because too high a gain may result in instability and a rapid oscillation around the set point (called *hunting*). This is because the larger gain causes the valve signal to be larger when there are small values of error, thereby causing the system to overreact to small changes in load. This overreaction causes an even larger error in the other direction, which again causes a large change in the valve signal, and we overshoot in the other direction. Adjusting the gain for stable control (minimal offset without hunting) is called tuning the control loop and is discussed further in Section 1.5.

A more common way of expressing the proportional gain is the term *throttling range*. The throttling range is the amount of change in the controlled variable that causes the controlled device to move from one extreme to the other, from full-open to full-closed. It is inversely proportional to the proportional gain. Applying integral logic along with proportional logic has the effect of minimizing or eliminating offset. This is called proportional plus integral (PI) control logic, expressed mathematically as:

$$V = V_0 + K_p e + K_i \int e \, dt \qquad (Equation\ 1\text{-}3)$$

The longer the error persists, the larger the integral term becomes, so that the effect is always to drive the value of the controlled variable toward the set point and eliminating offset. This is shown in *Figure 1-12*.

The sensitivity of the control logic is now a function of both the proportional gain K_p and the integral gain K_i. Just as with proportional control, it is possible to have unstable control if the gains are too high; they must be tuned for the application.

One disadvantage of including the integral term is an effect called *windup*. This is caused when the control loop is operating but the controlled device

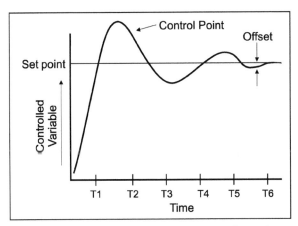

Figure 1-12   Proportional Plus Integral Control

is disconnected or otherwise not able to control the controlled variable, such as when a system is turned off at night. In this case, the controlled variable cannot be maintained at set point, so the integral term becomes larger and larger. When the system is turned on, the value of V is fully in one direction and the system will usually overshoot the set point. It takes time for the integral term to fall because of the long period that the system was far from set point. This effect (windup) causes the system to be temporarily unstable.

The problem can be mitigated by simply disabling the controller when the system is turned off (the preferred solution), by adding derivative control (discussed below), or by any number of anti-windup devices or algorithms commonly used with analog electronic controllers. A common anti-windup algorithm is to use proportional logic only until the system has been on for a period of time.

PI control is available with many pneumatic and analog electronic controls. It is virtually standard on digital control systems.

Using all three terms of *Equation 1-1* is called proportional plus integral plus derivative (PID) control logic. Adding the derivative term reduces overshooting. It has the effect of applying "brakes" to overreacting integral terms. Typically, derivative control has a very fast response, which makes it very useful in such applications as fast acting industrial processes and rocketry. However, because most HVAC system responses are relatively slow, the value of derivative control in most HVAC applications is minimal. Including the differential term may complicate the tuning process, and cause unstable responses. For these reasons, derivative control logic is *normally not used* in most field HVAC applications. (Note that the generic control loop, particularly in digital control applications, is often referred to as a PID loop, even though the derivative function is typically unused.)

While PID logic is generally applied to systems with continuous or modulating capacity capability, it may also be applied to systems with staged capacity capability to improve the accuracy versus two-position control logic. The way this is usually done is by applying a modulating control loop to the controlled variable, the output of which (V) is a "virtual" output (an output that does not actually control a real device). Then, step control logic is applied to a second control loop using the signal V as its controlled variable. The output of this second loop sequences the capacity stages of the equipment. Using PID logic in this manner can result in a smaller operating differential than using step control logic, particularly if the system has many steps of control (four or more).

Pulse-width Modulating, and Time-proportioning Control

Another type of modulating logic applied to an on–off type output is pulse-width modulation (PWM). The output is based on movement in a series of discrete steps, but it simulates true modulation quite well. Here, the output of the controller is a series of pulses of varying length (see *Figure 1-13*) that drive the controlled device (such as a stepping motor driving a valve or damper, or on–off control of an electric resistive heater). The output signal of the control loop (V) defines the length of the pulses rather than the position of the controlled device as it does with true modulating control. If the actual position of the controlled device must be known, as it may be for some control schemes, a feedback device that senses actuator position must be provided and fed back to the control system as another input.

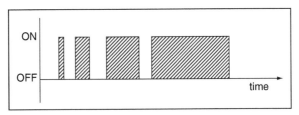

Figure 1-13   Pulse-width Modulation

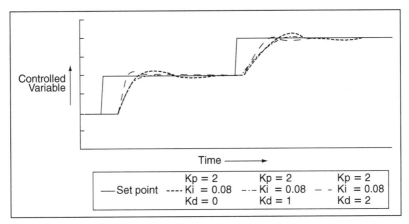

Figure 1-14   Time-proportioning Control

A variation on PWM is time-proportioning control (see *Figure 1-14*). Like PWM, the output is a series of on/off pulses, but the time cycle is fixed and the percentage of on-time and off-time during that cycle period is varied.

1.5 Gains and Loop Tuning

When you first started driving, chances are your actions were jerky, you applied too much gas when starting up, and overcompensated by releasing the accelerator too quickly or pressing too hard on the brakes. This is an example of an overly responsive control loop. As you gained experience driving, you effectively were tuning your control loop, subconsciously adjusting the sensitivity to which you responded to error, the difference between actual speed and desired speed. The adjustments depend to a certain extent on the car you are driving, the sensitivity of its accelerator and brakes, and the power of its engine. When you drive a different car, you must retune your driving control loop to adjust to these changes in system responsiveness.

HVAC system control loops must be similarly tuned. Every loop is a little different because the system to which it is applied is different. You can use the same controller to control the heating coil serving a small hotel room and to control a heating coil serving a huge warehouse. But the time constants of the two systems will be very different, so the controller gains must be adjusted, or tuned, to suit the two applications.

The effect of loop tuning can be seen in *Figures 1-15–1-17* which show, respectively, P-proportional, PI, and PID control loops with various gains.

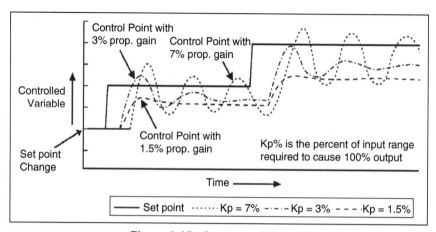

Figure 1-15 Proportional Control

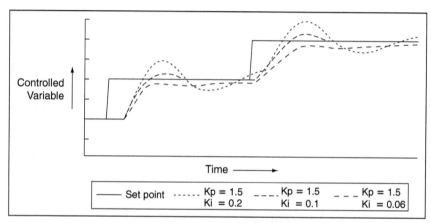

Figure 1-16 Proportional–Integral Control

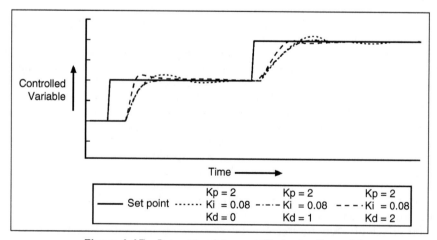

Figure 1-17 Proportional–Integral–Derivative Control

These figures show how the controller responds to step changes in set point. Except when a system is first started each day, it is unlikely in most HVAC applications to abruptly change set points in this way. But the response would be similar to large changes in heating and cooling loads, as might occur when an assembly room is quickly occupied, for instance. As can be seen by the curves, the accuracy and stability of the control can be optimized by selecting the proper proportional, integral, and derivative gains.

Loop tuning is currently somewhat of an art and is usually done empirically by trial-and-error. The technique is typically to tune the proportional gain first, then adjusting the integral gain to eliminate offset. (As noted above, the derivative gain is usually not used, at least partly because it complicates this tuning process.) The PID gains are initially set to values based on rules-of-thumb, manufacturer recommended values, or learned from experience with similar applications. The control technician will then observe the system in action and adjust the gain upward until oscillation is detected. If trend logging is available, the performance should be viewed over time. The gain adjustments will then be backed off to about one-half of the high value. (Note: defined performance of PID values may perform differently from one manufacturer to another.) The more experienced the technician is, the more precisely and more quickly the loop will be tuned. This procedure, while not optimum, will usually provide reasonable results.

More precise loop tuning techniques can be applied, but usually the process is too cumbersome to be done manually. Some digital control systems include automatic loop tuning software that applies these more rigorous loop-tuning techniques to automatically tune loops without input from the technician.

Proportional- or P-control logic assumes that processes are linear; the function that describes the error has the same characteristics independent of operating conditions. Most real processes are nonlinear and thus PID logic may be very difficult to set up to maintain zero error under all conditions.

For instance, when the valve serving the heating coil in *Figure 1-2* is opened only slightly, the supply temperature will rise very quickly. This is an inherent characteristic of steam or hot water heating coils. After the valve is opened 50%, opening it further has little impact on the supply air temperature. If we tune the loop to maintain excellent control when water (or steam) flow rates are high, it may be too sluggish and will provide poor control when water (or steam) flows are low. Conversely, if we tune the loop when flows are low, it may be too responsive and become unstable when flows are high.

To mitigate this problem, loops could be dynamically self-tuned, meaning the gains could be automatically and continuously adjusted to maintain precise control regardless of operating conditions. Some digital control systems have this capability and more systems are expected to as more "robust" (fast responding) self-tuning techniques are developed. Dynamic self-tuning also reduces commissioning time because it eliminates the need for manual tuning.

Another means to mitigate the nonlinearity problem is through the use of *fuzzy logic,* which is a relatively new alternative to PID control logic. Fuzzy logic imitates human intuitive thinking by using a series of fuzzy, almost intuitive, if–then rules to define control actions. Neural networks are another technique for self-tuning using artificial intelligence to "learn" how a system behaves under various conditions and the proper response to maintain control. More information on *fuzzy logic, artificial intelligence,* and *neural networks* can be obtained in more advanced control classes and texts.

1.6 Control Actions and Normal Position

Controllers may be direct acting (DA) or reverse acting (RA). These terms describe the control action or direction of the controller output signal relative to the direction of a change in the controlled variable. Direct acting means that the controller output increases as the value of the controlled variable increases. Reverse acting means that the controller output decreases as the value of the controlled variable increases. For example, a cooling valve is to control discharge air temperature at a set point. If the discharge air temperature is above set point the valve signal may be increased to open the valve to allow more cold water to flow through the coil. We would call this "direct acting," as when the temperature being controlled is above set point, we would increase the signal to the valve; we would say "temperature up, signal up," which depicts direct acting. Similarly, controlling heat is typically reverse acting: if the temperature was below set point, the signal would be increased or turned on. We would say "temperature down, signal up."

The term *control action* must be used with care because it is used in practice to describe many different control system characteristics. The term is used most commonly as above to describe the direction of the output signal relative to the direction of change of the controlled variable. But it is also frequently used to describe what is termed here as the control mode (for example, two-position action, floating action, modulating, etc.). It can also be used to describe the type of control logic used for modulating control (for example, proportional action, integral action, etc.).

Figure 1-18 shows how direct acting and reverse acting signals look using proportional control logic. The signal varies in direct proportion to the error signal, as described mathematically in *Equation 1-2*. The magnitude of the slope of the line is the proportional gain while the sign of the slope is positive for direct acting and negative for reverse acting. The throttling range is the

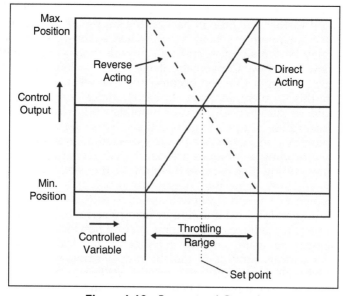

Figure 1-18   Proportional Control

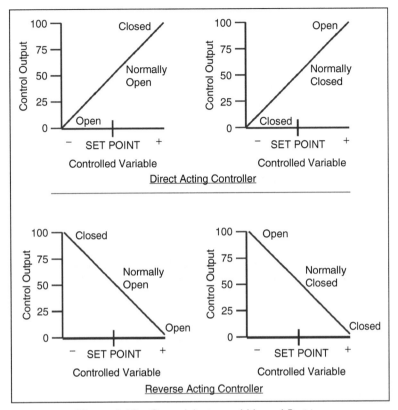

Figure 1-19   Control Action and Normal Position

difference in the value of the controlled variable to cause the controlled device to go from full-open to full-closed (see also *Figure 1-19*). Note how the set point is depicted as being in the center of the throttling range rather than at one extreme, as we depicted it for two-position and floating controls (*Figures 1-4–1-8*). This is a typical representation of proportional control logic. However, in practice, proportional controllers can be calibrated so that the set point (as it is set using the set point adjustment knob on the controller) is represented by any point within the throttling range (for example, that corresponding to fully open, fully closed, or any place in between).

Controlled devices (such as dampers, valves, and switches) may be normally open (NO) to flow through the process plant or normally closed (NC) to flow. These terms describe the so-called normal position of the damper, valve, or contact, which is the position it assumes when connected to its actuator but with no power (electricity or control air) applied.

Devices with normal positions must have some self-powered means of actuation, generally a spring built into the actuator or relay solenoid. The spring closes or opens the device when control power is removed. For example, a normally closed damper is one that is configured so that the spring in the actuator automatically closes the damper when the power to the actuator is removed or shut-off. If the actuator has no spring to return the damper to its normal position, the damper will simply stay in the last position it was in before power was removed. This type of damper/actuator does not have a normal position. The spring power must be large enough to do the job of returning the device to the intended position.

With three-way valves (valves that divert a stream of water or compressed air into two streams, or that mix two streams into one), one port is called the common port (the entering port for the diverting valve, or the leaving port for the mixing valve – the port which has continuous flow). The common port is open to the normally open port and closed to the normally closed port when control power to the valve actuator is removed. Whether a three-way valve is normally open or normally closed to flow through the controlled device depends on how the three ports are piped. (See Chapter 3 for schematics and more discussion of control valves.)

It is important to note that the normal position of a controlled device does not refer to its position during normal (everyday) operation. For instance, an outdoor air damper may be configured to be normally closed when in fact it is usually open during normal fan operation. The term *normal* here strictly refers to the position when control power is removed.

The use of spring-return actuators and other devices that have a specific normal position can be used to return the system to a fail-safe position should control power fail. For instance, hot water valves on outdoor air intake coils are typically configured to be normally open to the coil so that if control power fails, full hot water flow will go through the coil, thus preventing coil freezing or freezing of elements downstream.

The normal position can also be used as a convenient means to affect a control strategy. For instance, the inlet guide vanes on a supply fan and the outdoor air intake dampers to a fan system may be configured to be normally closed with the power source to the actuator or controller interlocked to the supply fan. (The term *interlocked* here means the power source is shut off when the supply fan is off and vice versa.) Thus, when the supply fan shuts off at night, the inlet guide vanes and the outdoor air damper will automatically close. This saves the trouble of adding controls that would actively shut the inlet guide vanes and outdoor air dampers when the fan turns off, as would be required if actuators without spring-return were used.

In most cases, the spring-return affects how the controlled device responds to a control signal. In general, when the control signal is reduced or removed (zeroed), the device moves towards its normal position. An increase in the control signal will cause the device to move away from its normal position. For this reason, the normal position must be coordinated with the control action of the controller and the nature of the process plant.

For example, in the heating system shown in *Figure 1-2*, if the valve is configured normally open, the controller (and thermostat) must be direct acting. This is because the valve will start to move toward its normal position as the control signal from the controller reduces. To close the valve, the control signal must be increased to its maximum value. Thus, as the air temperature in the duct rises, we want the control signal to rise as well so that the valve will close and reduce the amount of hot water passing through the coil, preventing the air from overheating.

If we replaced the heating coil in *Figure 1-2*, with a chilled water-cooling coil and the same normally open control valve, the controller would have to be reverse acting. This is because as the temperature of the air rises, we want the valve to open to increase the flow of chilled water. For the valve to open,

Table 1-3 Required Control Action

Application and Controlled Device	Normal Position	
	NO	NC
Heating valve or damper	DA	RA
Cooling valve or damper	RA	DA

the control signal must fall. Because this is the opposite direction of the temperature change, a reverse acting controller is required.

The relationship of control action to normal position for heating and cooling applications is shown in Table 1-3.

These relationships are also shown schematically in *Figure 1-19* for proportional controls.

In most applications, you would first select the desired normal position for the device based on what might be perceived as the fail-safe position, then select the controller direction (reverse acting or direct acting) that suits the normal position selected and the nature of the process plant (for example, whether the system is to provide heating or cooling). If there is a conflict (for example, you would like to use a normally open heating valve with a reverse-acting controller), a reversing relay may be added to change the action of the controller. Reversing relays (discussed in Chapter 8) have the effect of reversing the control action of the controller.

1.7 Control Range and Sequencing

The output from a control loop, V in *Equations 1-1–1-3*, can be scaled to yield values in a given range by adjusting the gains relative to the error. For instance, we may want a control loop output to range from 0% to 100%, with 0% corresponding to a fully closed controlled device for a direct-acting loop (or fully open for a reverse-acting loop) and 100% corresponding to full-open (full-closed for reverse acting). As we shall see in later chapters, the output from a pneumatic controller generally ranges from 3 psi to 13 psi while an electronic controller output can range from 2 Vdc to 12 Vdc.

For a controlled device to work with a given controller, it must operate over the same control range as the controller output, or a subset of that range. The control range is the range of control signal over which a controlled device will physically respond. For instance, for a pneumatic controller with a 3 to 13 psi output range, the controlled device must have a control range within that output range. Typical control ranges for pneumatic devices are 3 to 8 psi, and 8 to 13 psi. At each end of the control range, the device is fully in one direction (for example, fully open or fully closed for a valve or damper). For instance, a normally open pneumatic control valve with a control range of 3 to 8 psi will be fully open at 3 psi and fully closed at 8 psi. The control span is the difference between the signals corresponding to the extremes of the control range. In this pneumatic control valve example, the control span is 5 psi (8 psi minus 3 psi).

By properly selecting controlled devices with the proper control range and proper normal position, devices can be sequenced using a single controller.

Sequencing means that one device is taken from one extreme of its control range to the other before doing the same with the next device.

Energy codes typically require that the valves have a control range that is nonoverlapping. In this case, we would select the heating valve to have a control range of 0%–50%, and a cooling valve to operate from 50% to 100%. This would cause the heating valve to go from full open to full closed when the controller output goes from 0% to 50%, then from 50% to 100%, the chilled water valve to go from full closed to full open. The ASHRAE Standard 90.1-2004 requirement is even more stringent than just not overlapping:

> 6.4.3.1.2 Dead Band. Where used to control both heating and cooling, zone thermostatic control shall be capable of providing a temperature range or dead band of at least 5°F within which the supply of heating and cooling energy to the zone is shut off or reduced to a minimum.

To provide this 5°F requirement one would normally use a different strategy which will be covered later in the course.

For instance, suppose we wish to sequence a heating and cooling valve to control supply air temperature using a direct-acting controller with an output that ranges from 0% to 100% (see *Figure 1-20*).

The normal position of the chilled water valve must be opposite that of the hot water valve. This will cause one to open as the other closes using the same control signal. In this case, because the controller is direct-acting, the hot water valve must be normally open while the cooling coil valve must be normally closed. If, for any reason, we also wanted the cooling valve to be normally open as well, we could do so provided we also add a reversing relay between the controller and the valve. But this adds to cost and complication, and generally should be avoided if possible. Another way to accomplish this is with an electronic actuator which has a switch that allows the actuator to drive clockwise or counterclockwise as the control signal increases depending on the position of the switch.

1.8 Controls Documentation, Maintenance, and Operations

All control systems should be documented with written sequences, parts lists, data sheets, damper and valve schedules, control drawings, marked site plan locations for all controls and remote devices, and points lists where applicable. This documentation must be kept in a safe and accessible place for the life of the systems and building. After the installation is complete, total documented performance testing and commissioning, clean up and fire caulking of all penetrations should be made. Training sessions should be set up for all affected parties to participate in. All control systems require periodic maintenance, adjustments, calibration checks, and testing in order to stay in operation properly. Most often, this should be performed on a quarterly or semi-annual basis at minimum. Some standards have specific maintenance requirements such as ASHRAE Standard 62.1-2004 requirement that outdoor air dampers and actuators to be visually inspected or remotely monitored every 3 months. Records of all repairs and maintenance should be kept. You should refer to ASHRAE Standard 180 (to be published) for further details on this subject.

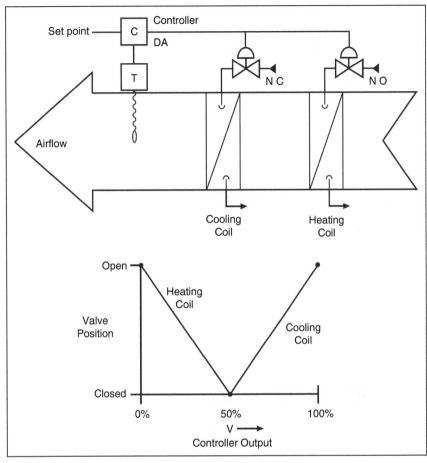

Figure 1-20   Sequencing

The Next Step

In the next chapter, we will learn about electricity fundamentals. Relays, transformers, motor starters, and other electrical devices play a role in almost all control systems, so it is important to understand the fundamentals of electrical circuits before going into detail about specific control hardware.

Bibliography

Donaldson, B., and Nagengast, B. (1994) *Heat and Cold: Mastering the Great Indoors.* Atlanta, GA: ASHRAE.

Harriman, L., Brundrett, G., and Kittler, R. (2001) *Humidity Control Design Guide.* Atlanta, GA: ASHRAE.

Chapter 2

Basics of Electricity

Contents of Chapter 2

Study Objectives of Chapter 2

This chapter introduces simple electrical circuits and common devices used to provide and control electrical power in HVAC systems. It is not intended to be a comprehensive course in electrical engineering, nor does it address electronics. (An understanding of electronics is important if the internal operation of analog electronic controllers and microprocessors is to be understood, but generally it is not necessary for most control applications.) This chapter is included because virtually all HVAC control systems will have relays, transformers, starters, and other electrical devices as a part of them. Many control systems are composed almost entirely of relays, so it is important to understand relay logic and how to read ladder diagrams of electrical devices.

After studying this chapter, you should be able to:

Understand basic electricity concepts and simple electrical circuits.
Understand the mathematical relationships between power, voltage, current, and resistance.
Understand how a relay works, why relays are used, and the symbols used to show normally open and normally closed contacts.
Understand how transformers work and why they are used in power services and control systems.
Understand what motor starters are and where they are used.

Understand why variable speed drives are used and the principles upon which many of them work.

Become familiar with Boolean logic and relay logic.

Learn to read ladder diagrams.

2.1 Simple Circuits and Ohm's Law

Electrical force is one of the primary forces interconnecting matter. Elementary particles such as protons and electrons have electrical charges. Particles of like charge repel each other while those with opposite charges attract. If two bodies of opposite charge are held apart, an electrical potential is created between them. If the bodies are then interconnected by material, loosely bound electrons in the material will be caused to move; a motion we call an electric current. Some materials (such as metals like copper) are more conducive to electrical flow. They are called conductors. Materials such as ceramics, rubber, and plastics offer a high resistance to free electron movement; they are called insulators.

Figure 2-1 shows a simple electrical circuit.

The battery creates an electrical potential across it, also called an electromotive force (EMF) or voltage *V*. The wires connect the battery to the device we wish to power, called the load, which has a resistance *R* to electrical flow. When the switch shown is closed (when it connects the wires entering and leaving the switch together), an electrical circuit is formed. The result is that electrical current *I*, will flow around the circuit. The magnitude of the current will be a function of the voltage *V* and the resistance *R*. The battery provides the electrical potential energy to move the electrical current through the resistance.

The circuit might be understood better by using an analogy. *Figure 2-2* shows a water tank with a pipe connecting it to the ground.

The height of the water tank is analogous to the electrical voltage; the higher the tank, the higher its "voltage" or energy potential. Gravitational force rather than electrical force is the source of the energy potential. Water flow

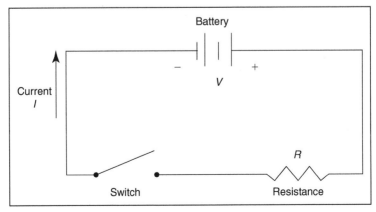

Figure 2-1   Simple Electrical Circuit

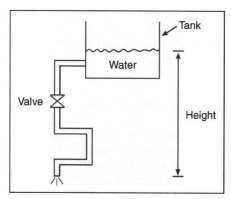

Figure 2-2   Water Tank Analogy

is analogous to electrical current and the shut-off valve is analogous to the switch. When the valve is closed, no water flows. When the valve is opened, water will flow through the pipe down to the ground due to the force of gravity. The rate of water flow (the "current") is a function of the height of the tank (the "voltage") and the frictional resistance to flow in the pipe, analogous to the resistance to electricity flow in our electrical circuit.

In this analogy, we close the valve increasing the resistance to stop the flow of water while we open the switch increasing the resistance to stop the flow of electricity. We will come across the same process when dealing with water and pneumatic control air valves and switches in the following chapters. Just remember that an open electrical switch allows no current to flow while a closed hydronic or pneumatic valve allows no fluid to pass.

The mathematical relationship relating the current, voltage, and resistance is expressed as:

$$V = IR \qquad\qquad (Equation\ 2\text{-}1)$$

where V is the voltage (in volts), I is the current (in amperes, or amps for short), and R is the resistance (in ohms, abbreviated Ω). This is called Ohm's law, named after George S. Ohm who was the first to express the relationship in 1826. Sometimes the letter E or e is used to represent the electrical potential instead of V. Others use E to represent electrical potential across a battery, and use V to represent voltage drop across a load. In this text, V is used for both purposes, which is common in general engineering practice.

For a practical example, often used in the field, if you wanted to convert a 0–20 mA signal to be 0–10 VDC, then you would put a 500 ohm resistor across the circuit, hence the new voltage would equal the 0.02 A times the 500 ohms which is 10 VDC.

You can imagine that if you added another pipe from the tank to the ground in our water tank example, you would still get the same flow rate through the first pipe no matter whether the valve through the second was opened or closed (see *Figure 2-3*).

This is called parallel flow because the two pipes are in parallel with each other (side by side). The same is true with electricity. The electrical potential

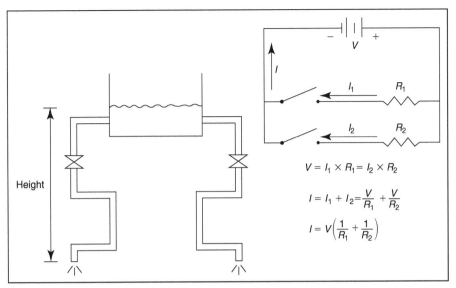

Figure 2-3   Parallel Circuits

across each resistor in *Figure 2-3* is still the same, V. Therefore, from Ohm's law we know that the current must still be the same:

$$I = \frac{V}{R} \qquad\qquad (Equation\ 2\text{-}2)$$

Of course, the total current from the battery is now increased. Just as our water tank will drain faster when we add the second pipe, our battery will "drain faster," or use more current, when there are two parallel circuits. If the resistances (R_1 and R_2) in *Figure 2-3* were equal we would get twice the current (flow).

$$I = \frac{V}{R} + \frac{V}{R} = \frac{2V}{R} \qquad\qquad (Equation\ 2\text{-}3)$$

If we took that same added piping and placed it in series with (added it to the end of) the original piping, the water would have to follow a longer and more circuitous route from the tank to the ground, see *Figure 2-4*.

We can imagine then that the water flow rate would go down because we increased the resistance to flow. Again, the same is true in the analogous electric circuit. The electrical potential is split between the two loads (resistors), so the current must fall. We know that the voltage drop across the first resistor plus that across the second must add up to the total battery voltage. Because the voltage drop must be equal to $I \times R$ from Ohm's law and the current through each resistor must be the same (just as the water flow through the two pipe sections in series must be the same), and if the resistances (R_1 and R_2) in *Figure 2-3* were equal we would get ½ the current (flow).

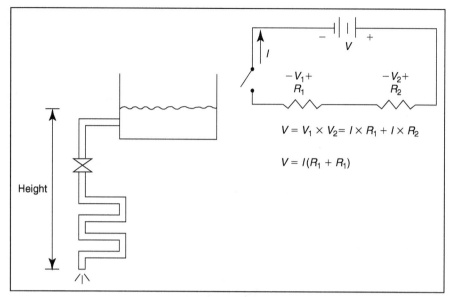

Figure 2-4   Series Circuits

$$I = \frac{V}{R+R} = \frac{V}{2 \times R} \qquad \qquad (Equation\ 2\text{-}4)$$

So the current is cut in half when we put the two identical resistance loads in series. The amount of overall potential energy stored in our water tank is a function of the amount of water in the tank and its height above the ground. If we double its height, we double the amount of stored energy because now we can get twice as much flow through our pipe. If we double the volume of water in the tank, we also double the energy stored because it will last twice as long once we start to drain it. So the energy stored is then the product of the height and the volume of water.

The rate at which we discharge the energy (called the power) is then the product of the height and the rate at which water flows out of the tank. Using our analogous electrical terms, electrical power (P) in watts is then:

$$P = IV \qquad \qquad (Equation\ 2\text{-}5)$$

Using Ohm's law to substitute for $V = IR$, this can also be expressed as:

$$P = I \times R \times I$$
$$P = I^2 \times R \qquad \qquad (Equation\ 2\text{-}6)$$

2.2 AC Circuits

In the previous section, we looked at a simple circuit using a battery maintaining a constant electrical potential or voltage that resulted in a constant current through the circuit. This is referred to as a direct current (dc) circuit because the current is constant and flows only in one direction. But the electrical systems serving our homes and buildings are not dc systems. Rather, they are alternating current (ac) systems.

Alternating current is produced by varying the electrical potential (the voltage) in a sinusoidal fashion, as shown in *Figure 2-5*. The voltage varies from positive (+V) to negative (−V). In the simple circuit shown in *Figure 2-6*, the alternating voltage causes a corresponding alternating current that swings back and forth in the same sinusoidal fashion as the voltage. When you measure the ac voltage you measure the "root mean square," rms voltage which for a pure sine wave is 0.707 times the peak voltage.

The time it takes for the oscillation to go through a full cycle (from peak to peak) is called the cycle time, measured in cycles per second or Hertz (Hz). Common ac power sources in the United States are 60 Hz, while those in many other countries are 50 Hz.

Ohm's law and the formula for power derived for dc circuits in the previous section also apply to the simple ac circuit shown in *Figure 2-6*. However,

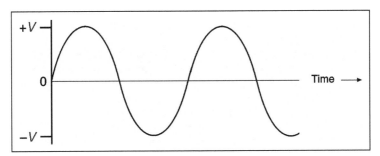

Figure 2-5   Alternating Voltage

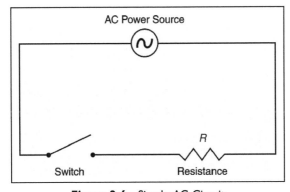

Figure 2-6   Simple AC Circuit

ac circuits are never so simple in real applications. The rms current in an ac circuit with resistance is given by the rms voltage divided by the resistance ("root mean square voltage" means the effective average voltage). Two other elements – capacitance and inductance – come into play, which make ac circuits more complicated.

In the circuit shown in *Figure 2-7*, we have a capacitor in series with a resistor. A capacitor, as its symbol implies, is two parallel plates typically separated by a nonconducting material called a dielectric. If we applied a dc (constant) voltage across the capacitor by closing the switch, the current would start as it would if the capacitor was not in the circuit, as electrons begin to move due to the electrical potential from the battery. However, as electrons collect on the plates of the capacitor, they begin to repel other electrons and the flow diminishes. This is shown in the graph on the right of *Figure 2-7*.

Quickly, the current would drop to near zero, actually a very low amount called *leakage* due to the very low but still non-zero conductivity of the dielectric. At this point, energy is stored in the capacitor, and is fully charged. The capacitor can be thought of as a storage device. We measure its size using the term capacitance *(C)*, which is a function of the size of the plates, their separation distance, and the properties of the dielectric.

As *Figure 2-7* indicates, a capacitor is like an open switch in a dc circuit once steady-state is reached, which is why it does not play a part in most dc systems. But when a capacitor is placed in an ac circuit, the result is very different. The alternating voltage simply allows the capacitor to charge, then discharge, then charge again. While no electrons actually pass through the capacitor, there is still current flow back and forth through the load just as it would if the capacitor was not there.

The only difference is that the capacitor causes a time delay in the current. It takes time to charge and discharge the capacitor plate, so current and the voltage become out of phase, as shown in *Figure 2-8*. Another effect that has an impact on ac circuit current flow is inductance, which is a resistance to current change caused by magnetic induction. As current flows through a wire, a magnetic field is created around it, as shown in *Figure 2-9* (with current directional arrows shown as the direction electrons follow, flowing from positive to negative).

If we wound the wire into a spiral, the magnetic fields from the wires would align inside the spiral, creating a strong magnetic field in the center.

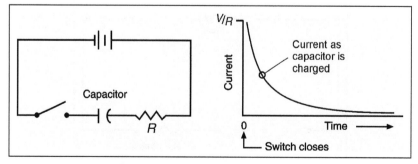

Figure 2-7   Capacitor in DC Circuit

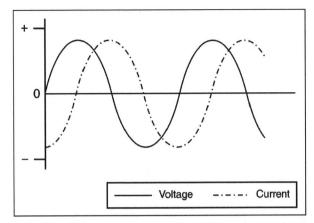

Figure 2-8   Voltage and Current Out of Phase, Current Lagging

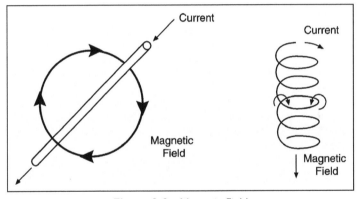

Figure 2-9   Magnetic Field

This principle is used in many electrical devices (such as solenoids, relays, and electromagnets).

The interaction goes the other way as well; if a changing magnetic field passes through a conductor, it induces an electromotive force perpendicular to the magnetic field. This phenomenon, discovered by Hans Oersted in 1820, can affect the current in a circuit, as can be seen in *Figure 2-10*. When the switch is closed, current begins to flow through the circuit. The rising (changing) magnetic field in the coil (the inductor) creates an electrical potential that is opposite to that of the battery. This potential, called back-EMF, cancels part of the potential of the battery, effectively retarding flow in the circuit. Eventually, as the current reaches steady-state, the back-EMF begins to die away because the current is no longer changing. Once steady-state is reached, the inductor has no impact on the dc circuit, as can be seen in *Figure 2-10*.

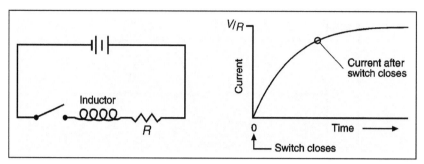

Figure 2-10   Inductor in DC Circuit

When the switch opens the current flow will abruptly stop. A large EMF (voltage) will be generated as the energy stored in the magnetic field in the coil starts to collapse. Depending on the size of the current flowing before the switch was opened, this can lead to arcing (sparking) across the contacts and generating electromagnetic interference (EMI) (electrical radiated energy), which can affect sensitive electronics.

Things are very different in ac circuits because the current is always changing. The back-EMF always acts to retard the change in current whether the change is an increase or a decrease. The ability of the inductor to resist changes in current is measured in terms of its inductance L.

The impact of inductance on current flow is to push the voltage and current oscillations out of phase with each other, like the capacitor did. However, while the capacitor causes current to lead voltage, the inductor causes the current to lag voltage, as can be seen by the graphs on the right of *Figures 2-7* and *2-10*. The effects are opposite and in fact may tend to cancel each other.

The capacitance and inductance effects on current flow are collected into a factor called impedance, Z, which is analogous to resistance in a dc circuit:

$$V = I \times Z \hspace{4cm} (Equation\ 2\text{-}7)$$

Capacitance and inductance effects exist even where we do not want them to. For example, alternating current running in one circuit will induce current in other adjacent circuits, and poorly made wiring connections will create a capacitance between the contacts. The fact that these effects are always present is why there are no simple and purely resistive ac circuits, such as the one shown in *Figure 2-6*.

The power used in ac circuits is computed like that for dc circuits: it is equal to the product of voltage times the corresponding current. However, when the current and voltage are out of phase, as in *Figure 2-7*, the peak voltage and peak current do not occur at the same time. We can still relate the power to the rms current and rms voltage, but a factor must be added to account for the phase offset. This factor, called the power factor, varies from 0 (180° out of phase) to 1 (in phase):

$$P = V_{rms} \times I_{rms} \times PF \hspace{3cm} (Equation\ 2\text{-}8)$$

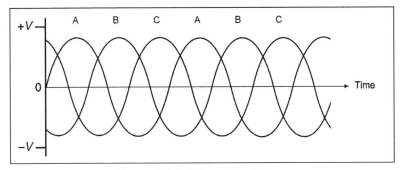

Figure 2-11   Three-phase Voltage

Because power plants, transmission lines, transformers, and other distribution equipment have to be sized for the maximum current, utilities often impose an added charge to users that have very low power factors. A low power factor is typically caused by an inductive device such as a motor. Capacitors, which as noted above have an opposite effect on current/voltage phase shifts as inductors, can be used to correct power factor in these instances.

Most larger buildings are served by three-phase power services (discussed in the next section), typically with four wires, three of which are hot and one that is neutral (grounded at the service entrance or distribution transformer). The three hot wires have ac voltages that are 120° out of phase with each other, as shown in *Figure 2-11*. Three-phase (abbreviated to 3Ø) power is used because it is convenient for motors (it causes a naturally rotating field that starts a motor turning in the right direction) and it reduces the amount of current required for the same amount of power. The power equation for a three-phase circuit is:

$$P = V_{\text{rms}} \times I_{\text{rms}} \times PF \times \sqrt{3} \qquad\qquad (Equation\ 2\text{-}9)$$

2.3 Transformers and Power Services

A power transformer is a device that is used primarily to convert one ac voltage to another. Based on the principle that alternating current in an inductor electromagnetically induces an electrical potential in adjacent conductors, transformers can interconnect two circuits without any electrical connection between them. A transformer is shown schematically in *Figure 2-12*. It is essentially composed of coiled wiring (inductors) in each circuit which is connected by an electromagnetic field. Current flow in the primary circuit induces a voltage across the secondary coil.

If the number of coils in the primary circuit differs from the secondary, then the induced voltage will differ from the primary. The secondary voltage will be higher or lower by the ratio of the number of turns in each system. For instance, if there are four times as many windings on the primary side of the transformer as there are on the secondary, the voltage on the secondary

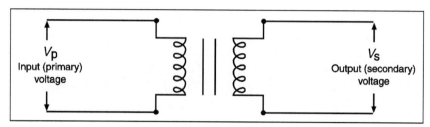

Figure 2-12   Transformer

will be ¼ that on the primary side. By varying winding ratios, transformers can be used to boost or reduce the voltage in the secondary circuit. If it boosts the voltage it is called a step-up transformer; if it drops the voltage it is called a step-down transformer.

The transformer's capacity is rated in VA (volt-amps) or kVA (kilovolt-amps), which are roughly equal to the product of the voltage times the maximum current required on either side of the transformer. The primary and secondary voltages are specified as rms voltage. This will be larger than the power (wattage) requirement because of the power factor shown in *Equation 2-9*.

The ease with which voltage can be boosted and reduced with ac systems is the main reason they are used for power distribution instead of dc power. The higher the voltage, the lower the current is for the same power output. There-fore, with ac systems, power from power plants can be transmitted to the communities they serve at very high voltages (low current), thereby allowing the use of smaller wiring and reducing power losses in transmission lines. This reduced current at high voltages is very significant as the power loss in transmission is proportional (*Equation 2-6*) to the square of the current $P = I^2 R$. Doubling the voltage and halving the current provides the same power but the losses are reduced to $.5 \times .5 = .25$ or 25%. The voltage can then be reduced using a transformer at the building service entrance to a lower, safer voltage to power lights and motors, and all the other power consuming devices in the building.

Three-phase transformers have three secondary, or output, windings. These windings may be connected in two ways – Delta or Wye Figure 2-13. In the Delta connection, the ends of the three windings are connected together and the same voltage is available across each phase. In the Wye connection, the three phases are all connected at one end: a common point. This point, called common or neutral, is usually grounded. The Wye connection provides a phase-to-phase voltage and a phase-to-common $1/\sqrt{3}$ times the phase-to-phase voltage.

The typical commercial building is provided (downstream of the utility transformer) with either a 208 V, 230 V, 480 V (or other), three-phase, four-wire service from which various voltages can be obtained depending on which wires are used. The voltages available with two of these service types are shown in *Figure 2-14*. When there is a 480-V service, there typically will be one or more transformers in the building that step down the voltage to 120 V, which is the voltage used by most common household and office plug-in appliances in North America.

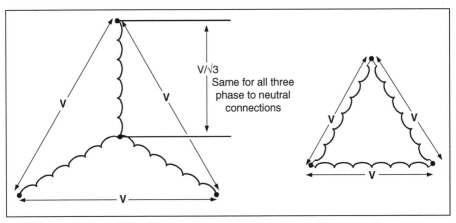

Figure 2-13   Delta and Wye Connected Three-phase Transformer Outputs

The lowest voltage available with typical services is still higher than that needed for typical control systems. Control systems use very little power, so the cost advantages offered by higher voltages (lower current, smaller wire sizes) are negligible. Therefore, control systems typically are powered by another step-down transformer producing 24 V. This is sufficiently low that

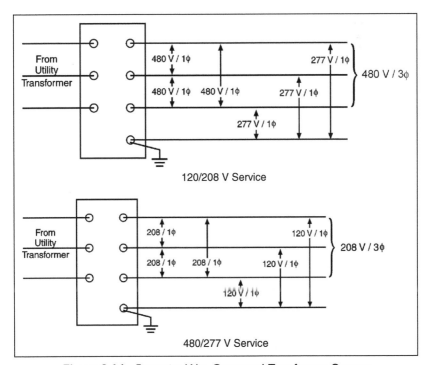

Figure 2-14   Four-wire Wye Connected Transformer Outputs

most building codes do not require many of the expensive protection devices (such as conduit and junction boxes) that are required for higher voltage circuits. This reduces installation costs. Typically, control transformers are small (less than 75 VA), which places them in a lower cost electrical code category (NEC Class 2, which includes power systems rated at less than 30 V and 100 VA).

2.4 Relays

A *relay* is an electrical *switch* that opens and closes under control of another electrical circuit. The electromagnetic relay with one or more set of contacts was invented by Joseph Henry in 1835. There are many types of relays used in controls, which will be discussed in later chapters. In simple electrical circuits, the term refers to a device that delivers an on/off control signal. In a sense, the relay is a type of remote-controlled switch.

Figure 2-15 shows a type of electromagnetic relay. It works by taking advantage of the magnetic force that results from current flow. Wire is wound around an iron core. When current is allowed to pass through the wire in either direction (when a voltage is placed across leads labeled A and B in the diagram), a magnetic field is formed in the iron. Iron is highly permeable to magnetic fields; much like copper is highly conductive to current flow. The result is that an electromagnet is created. The magnet attracts the metal armature (sometimes called a clapper), which is allowed to move toward the magnet, causing the contacts to close (connect to each other), connecting leads labeled C and D in the schematic. The contacts are often called "dry" contacts because they are electrically isolated from other circuits, including the circuit that energizes the relay, and hence they have no source of voltage in and of themselves. The relay solenoid coil is represented as a circle while the relay contact is represented in wiring diagrams by two parallel lines.

Figure 2-16 shows two simple circuits with a relay interconnecting the two. When the manual switch in the first circuit is closed, the relay is energized,

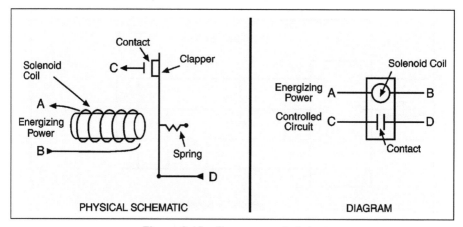

Figure 2-15   Electromagnetic Relay

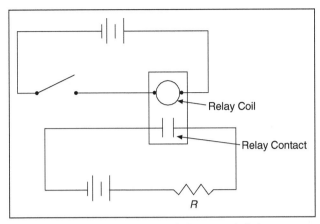

Figure 2-16   Simple DC Circuit with Relay

which closes the relay contacts in the other circuit. In this way, the switch clo-sure "information" was transferred from the first circuit to the second circuit without requiring that the two be electrically interconnected. Because they are separate, the voltage in the first circuit may be different from the voltage in the second, or the power may be from different sources. This is a common rea-son for using a relay.

Relays that have more than one relay contact are called multipole relays. Each pole (contact) is electrically isolated from the other, so each may be used to control a different circuit. Each pole is energized and de-energized at the same time, and they work in unison.

Multipole relays are represented in two ways. For simple diagrams, the relay contacts may be grouped under the relay coil so it is clear the contacts are controlled by that coil, as shown in *Figure 2-17*. When the drawing gets more complicated, this type of representation can crowd the drawings and make it confusing. In this case, the relays and contacts are labeled. For exam-ple, a three-pole relay R1 has contacts labeled R1-1, R1-2, and R1-3. In this

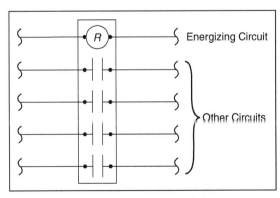

Figure 2-17   Multipole Relay

way, the contacts may appear anywhere in the diagram without having to show wires running back to the relay.

In Chapter 1, when speaking of controlled devices such as valves and dampers, the term *normal position* referred to the state of the device when control power was removed. For a general purpose relay, the *normal position* refers to the state of the contact when the relay is not energized. Thus, a normally open contact, N.O., is open when the relay is not energized (no current is allowed to flow through it) and closes when the relay is energized. Normally open contacts are shown in *Figures 2-15–2-17*. A normally closed contact, N.C., (represented diagrammatically as a contact with a slash through it) operates in an opposite manner to the normally open contact; it is closed when the relay is not energized and opens when the relay is energized. Normally open and normally closed contacts are shown in *Figure 2-18*.

Relay contacts may be single-throw or double-throw. A single-throw contact is much like a two-way control valve controlling a fluid, while the double-throw contact is much like a three-way valve. *Figure 2-18* shows both types of contacts used in simple circuits to turn on pilot lights. The contact on the double-throw relay has a common terminal that is simultaneously switched to the normally open terminal when the relay is energized and to the normally closed terminal when the relay is de-energized.

In *Figure 2-18*, when the manual switch is closed, the relay is energized, causing all the normally open contacts to close and all the normally closed contacts to open, turning on the red (R) pilot lights and turning off the green (G) pilot lights. When the switch is open, the relay is de-energized, closing the normally closed contacts and opening the normally open contacts, turning on the green pilot lights and turning off the red pilot lights.

Relays are commonly used as control logic devices and they are discussed in more detail in Chapter 4.

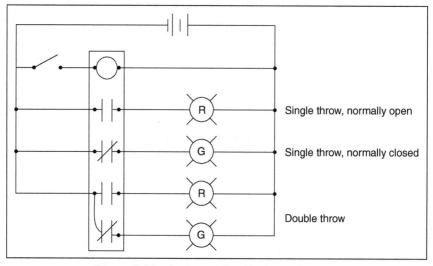

Figure 2-18   Single- and Double-throw Contacts

2.5 Motors and Motor Starters

Electric motors are devices that convert electric energy into mechanical energy by causing a shaft to rotate. The shaft can then be connected to a device such as a fan or pump to create useful work. The rotation in the shaft is caused by the interaction of two magnetic fields in the motor: one produced by the fixed part of the motor (the stator) and the other produced by the rotating part (the rotor) which is connected to the shaft. The magnetic fields are created either by fixed magnets or, more commonly, by an electromagnet created by a winding in which current flows. The most common type of motor is an ac induction motor. (A more complete discussion of motor types is beyond the scope of this course.)

Small motors used for HVAC applications tend to be powered using single-phase services, usually 120 V, 208 V, and 277 V in North America. Motors that are 5 horsepower (hp) and larger are typically three-phase motors, usually 208 V or 480 V.

Motors are generally controlled by a motor starter that serves two functions:

- *Start/stop control.* Motors require some means to connect power to them when they are required to operate, and to disconnect power when they are not. Starters may be manual (basically a large switch) or they may be automatic. The automatic type typically uses a magnetic contactor, using a concept similar to the electromagnetic relay described above. More recently, solid-state (electronic) starters have begun to be used, particularly on large motors such as those used on chillers.
- *Overload protection.* Electric motors are not inherently self-limiting devices. The device they are driving may operate in a fashion that causes the motor to draw more current than it was designed to handle. This is called overloading. There are other causes of high motor current such as motor insulation resistance breakdown, low voltage to the motor, voltage phase imbalance, etc. When high motor current occurs, the motor windings will overheat, damaging the insulation that protects the windings, resulting in motor burn-out. To protect against this, electrical codes generally require that motors be protected from thermal overload.

    A current sensing device called an overload relay provides overload protection. When the relay senses an overload condition, it drops out the motor contactor and the motor shuts off. A common overload device is a small electric resistance heater in series with the motor that warms a bimetallic strip in a manner similar to the warming of the windings in the motor caused by the same current. At high temperatures, the bimetallic strip warps and breaks a contact, tripping the motor. In *Figure 2-19*, the normal operating current is shown as just over 5 A. If the current rises to a steady current above 9 A the thermal overload protection will trip the motor off. The thermal overload takes time to respond just as the motor takes time to heat up, so the thermal overload trip does not trip due to the high, 25 A, startup inrush current.

However, there are situations such as a short circuit or a locked rotor (rotating part of motor is jammed and cannot rotate) when very high currents can flow and instant tripping is required. This is called over-current protection.

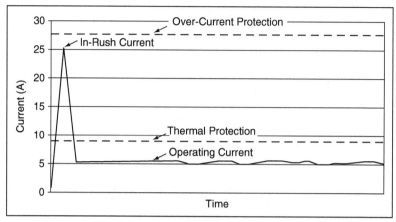

Figure 2-19   In-Rush Current and Overload Protection

The over-current devices (fuses, and circuit breakers) are primarily intended to protect against short-circuits. These devices must be set to trip at relatively high currents due to the in-rush current typical of motors when they are first started. This temporary in-rush current is much higher than the overload current. For this reason, over-current protection cannot protect against overload.

Some single-phase motors have internal thermal overload protection, a small switch that requires either manual or automatic reset, depending on the type used. For these motors, starters are not required and a simple line-voltage contact (such as a thermostat) may be used. All other motors will require starters, most commonly the automatic magnetic type.

An across-the-line magnetic starter for a three-phase motor is shown in *Figure 2-20*. To start the motor, a voltage is placed across the contactor coil (basically a relay device). Provided the overload relays are not tripped, the coil will energize and close the three contacts, connecting the incoming power to the motor. The motor will start operation in a direction. Always check for proper rotation of the motor in relation to what it is connected to, i.e. a fan blade or pump impeller: if the rotation is backward, simply swap two of the three-phase wires and re-terminate the contacts, and the motor direction will reverse.

The contactor often is fitted with one or more auxiliary contacts that are added poles that close and open along with the motor contacts. These can be used to interlock other equipment to this motor or to provide an indication that the motor has been started (lighting a remote pilot light, for instance). Make sure the voltage/amp ratings of the interlocked device will agree with the voltage going through those auxiliary contacts. For discussion purposes, often these auxiliary contacts are used to interlock other devices such as fans, pumps, dampers, and smoke dampers. Always be aware of the amperage ratings of these auxiliary contacts so that the number of interlocked devices you connect to them does not exceed their ratings for current, wire size, wire insulation value (600 V rating), and voltage, especially if you are using low voltage (24 V) over a long distance always size the wire appropriately.

Another concern is the minimum ampere rating of the auxiliary contacts. If the amperes are below the minimum rating, the contacts will not be cleaned

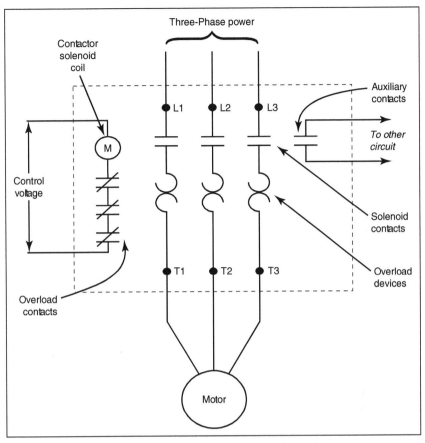

Figure 2-20 Across-the-line Magnetic Motor Starter

and erratic operation will occur. Remember, the wire insulation value, typically 600 V for commercial applications, must always match the highest value used if you are using it in conduit or in panels. In *Figure 2-20*, the control power comes from a source external to the starter, potentially a dedicated control circuit. It is better practice to draw the control power from within each starter itself. In that way, any interruption in the control circuit will only affect that one motor, whereas with a common control circuit serving many motors, a loss of control power would cause all of the motors to be inoperative.

Control power can be taken directly from two legs of the incoming power or one leg and the neutral. This is common with 208 V three-phase starters because 120 V can be achieved between any leg and the neutral (note: this is only true for Wye connected loads with the center common, or neutral). With 480 V starters, the resulting control voltage would be higher, and it is more common to see a control transformer mounted in the starter to step-down the voltage to 120 V or, less commonly, 24 V (see *Figure 2-21*).

The use of 24-V transformers would appear to be more economical because it would allow for less expensive installation of any remote control contacts.

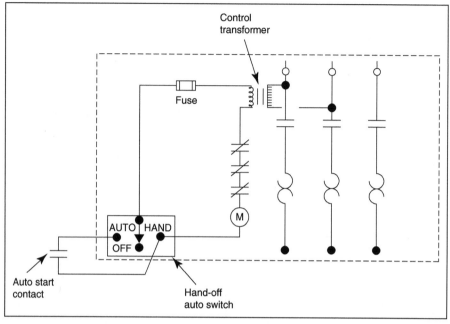

Figure 2-21   Magnetic Starter with H-O-A Switch

However, the amperage required to pull in the contactor at 24 V, termed VA, can be very high so the cable size is uneconomic and 120+ V is used. The wiring to any remote devices must be sufficiently sized to ensure that the voltage drop in the wiring does not result in an insufficient voltage at the contactor coil. To avoid this potential problem, many commonly practiced starter control transformers are 120–480 V.

Make note that the fire alarm start–stop relay, F/A, should always break the magnetic coil directly and before any other safety device: Life safety always takes first, and absolute, priority.

Also shown in *Figure 2-21* is a starter-mounted hand-off-auto (H-O-A) switch. The motor can be started manually by switching the H-O-A switch to the hand or on position. When in the auto position, the motor starts only when the remote contact closes. This might be a start/stop contact at an energy management system panel or simply a thermostat or other two-position controller.

There are many other types of starters in addition to the across-the-line starters shown in *Figures 2-20* and *2-21*. For instance, larger motors (larger than 75 hp to 100 hp on 480 V services) are typically served by reduced-voltage starters that are designed to "soft-start" the motor to limit the in-rush current. However, the controls system designer often does not know and seldom cares what type of starter or motor is used. Only the control circuit side of the starter is of importance to the design of the control system. For this reason, the starter is often depicted without any of the power-side wiring, such as in *Figure 2-22*. You may also note the different way the H-O-A switch is represented in *Figure 2-22* as compared to *Figure 2-21*. Both styles are commonly used.

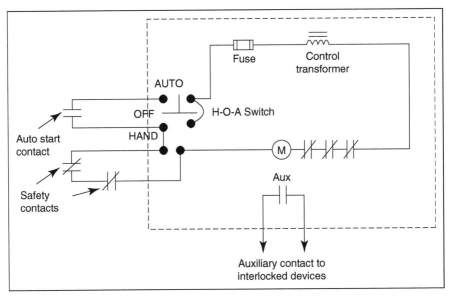

Figure 2-22   Generic Magnetic Starter with H-O-A Switch and Auxiliary Contact

Figure 2-22 includes wiring to safety enabling devices (such as a smoke detector, firestat, fire alarm device, or freeze-stat) that will stop the motor when either of the normally closed contacts open (upon detection of smoke or when freezing temperature air is sensed at a coil, for instance). These safety contacts are wired so that they can stop the motor whether the H-O-A switch is in either the hand or auto position.

Figure 2-23 shows a starter circuit using momentary contact pushbuttons to manually start and stop the motor. When the normally open start button is

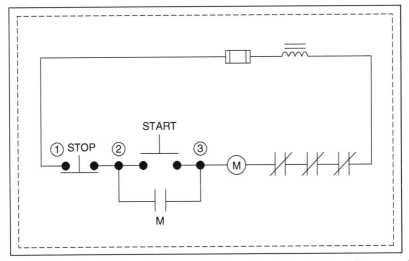

Figure 2-23   Generic Magnetic Starter with Manual Start/Stop Buttons (Three-wire)

depressed, the magnetic starter coil is energized, then the holding contact (an auxiliary contact of the starter coil) maintains power to the coil. When the normally closed stop button is pressed, power to the coil is removed, dropping out the holding contact so that when the button returns to its closed position, the motor stays off. This starter wiring is often called a three-wire control circuit because if the buttons were external to the starter, three wires would have to be run from the buttons to the starter, connecting at the points labeled 1, 2, and 3 in *Figure 2-23*. (The H-O-A switch approach shown in earlier diagrams is similarly called a two-wire control circuit because only two wiring connections would have to be made if the switch were remotely mounted.)

The two-wire (H-O-A) and three-wire (pushbutton) starter control circuits have two significant differences:

- First, if the motor was running and control power was interrupted and then resumed, a motor controlled by a two-wire circuit would restart. With the three-wire design, the holding coil would drop out when control power was lost, so the motor will only restart if the start button is pressed again.
- Second, the three-wire circuit can only be started and stopped using the pushbuttons. With the two-wire design, a single remote contact (such as a timeclock contact) can be used to start and stop the motor. For this reason, the two-wire approach is more common in HVAC control applications.

Sometimes an HVAC fan must serve double duty as part of the life safety system. For instance, in case of fire, a supply and return fan system may be used to exhaust smoke from one area of a building while pressurizing other areas to prevent smoke migration. The fan will generally be controlled by a fireman's control panel that is part of the life safety system, but a lot of times it is controlled from the fireman's panel as well as from the HVAC control system. For smoke control, most codes allow for the design engineer to design an "engineered smoke control" system that will allow for fans to operate during a fire condition. Typically, the HVAC control system controls it in the automatic side, and a manual switch and control is provided at the fireman's control panel location so that it can be overridden by the attending fire official on site. Typically, the HVAC control panel and fireman's panel are remotely located in a common location of the building. The design of these smoke control systems is complex and intricate, and much coordination is needed between the mechanical and electrical fields of expertise, both in design and during construction.

Figure 2-24 shows how these controls might be wired. The life safety system can command the fan to be off or on. These commands must take precedence over the automatic control of the system and any manual control at the H-O-A switch. When the off-relay shown is energized, the fan cannot run regardless of the H-O-A switch position or the demands of the automatic control system. When the on-relay is energized, the fan will run regardless of the H-O-A switch position and, the way this example is wired, regardless of the freeze-stat contact status. Wiring the freeze-stat in this manner allows the fireman to keep the fan on, even when the weather is freezing; this is valid control logic if life safety concerns are felt to override the concern about damaging

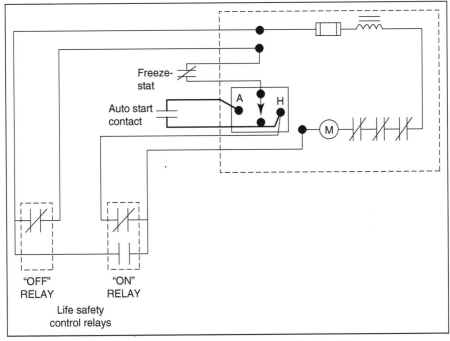

Figure 2-24 Magnetic Starter with Life Safety Fan Wiring

the cooling coil. The wiring between the starter and the fireman's control is generally provided by the life safety contractor, while the wiring connecting the starter to normal HVAC control systems, freeze stat and auto start contact in Figure 2-24, is usually provided by the controls contractor. The HVAC system designer must make this distinction clear in bid documents.

2.6 Variable Speed Drives

Variable speed drives (VSDs), also called adjustable speed drives (ASDs), are devices that can vary the speed of a normally fixed speed motor. In HVAC systems, they are used primarily to control fans in variable air volume systems instead of other devices such as inlet vanes and discharge dampers. Variable speed drives are more energy efficient than these other devices (their main advantage), but they also reduce noise generation at part-load, allow fans to operate at much lower loads without causing the fan to operate in surge (an unstable condition that can result in violent pulsations and possibly cause damage to the fan), and reduce wear on mechanical components such as belts and bearings. Variable speed drives are also used to control pumps on variable flow pumping systems and to control refrigeration compressors in centrifugal chillers.

Many types of variable speed drives have been used over the years, starting with dc drives used primarily in industrial applications, and mechanical drives that varied sheave diameter. One of the most important developments in recent years has been the advancement of variable frequency drive (VFD) technology. These drives use solid-state electronic circuitry to adjust the frequency and voltage of the power to the motor, which in turn varies the speed.

The most common VFDs used in HVAC applications are inverters using sine-coded pulse-width-modulation (PWM) technology. A schematic of the PWM is shown in *Figure 2-25*. The PWM works by first converting the incoming ac power to dc using a diode bridge rectifier. The voltage is then filtered, smoothed and passed on to the PWM inverting section. The inverter consists of high speed bipolar transistors that control both voltage and frequency sent to the motor.

The output, shown in *Figure 2-26*, consists of a series of short-duration voltage pulses. Output voltage is adjusted by changing the width and number of the voltage pulses while output frequency is varied by changing the length of the cycle. The waveform that is produced has the required voltage and frequency to produce the desired motor speed and torque, but it is not as smooth as the incoming sinusoidal source. For this reason, motors should be specifically chosen with adequate design and construction to withstand the less smooth power source.

Do you remember earlier in the chapter when we noted that when we measure the voltage and current in a pure sinewave ac circuit we are measuring the rms value? In most situations, being clear about rms does not matter as the waveform is close to a sinewave. As you can see in *Figure 2-26* the waveform is not a sinewave. A standard meter may read substantially high or low in this

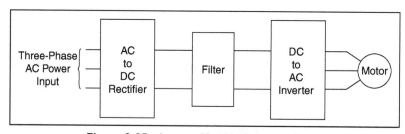

Figure 2-25   Inverter Variable Frequency Drive

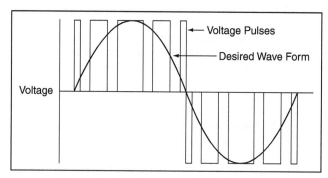

Figure 2-26   Sine-coded PWM Waveform

nonsinewave situation. With the output of a VFD it really matters that you use a meter designed to give you "true rms" readings. Meters are sold as "true rms" meters and are more costly than those which need the true sinewave input.

Variable speed drives (VSDs) take the place of a starter. They have both starting capability and overload protection built in. In fact, the microprocessor controls in most drives provide additional protection against other faults (such as under-voltage, over-voltage, ground-fault, loss of phase, etc.). Variable speed drives also provide for soft-start of the motor (if so programmed), reducing in-rush current and reducing wear on belts and sheaves.

While a starter is not required when a VSD/VFD is used, one may be provided as a back up to the drive so that the motor can be run at full speed in case the drive fails. A schematic of how a bypass starter might be wired is shown in *Figure 2-27*. Bypass starters were considered almost a requirement in the early days of VSDs and VFDs but now, as drives have improved in reliability, the need for bypass starters is a lot less critical. If a bypass is required, sometimes the use of multiple drives being fed from one similarly sized bypass is desirable, and reduces the cost of buying multiple bypasses. When using a bypass starter, it is important to consider how well the system will operate at full speed. For instance, in a VAV fan application, operating the fan at full speed may cause very high duct pressures at low airflow rates, potentially damaging the duct system. Some new VSDs have what are called electronic bypasses, which are speed selectable and do not have to run at full speed. (These electronic bypasses are not independent; therefore, they use the same contactors and overloads, as does the VSD, so they are not totally

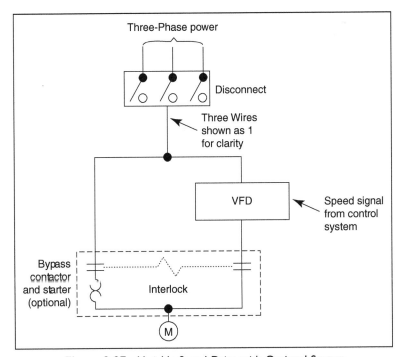

Figure 2-27   Variable Speed Drive with Optional Starter

independent.) Some other means of relieving the air pressure or reducing fan speed should be provided. These complications, along with the added costs involved, must be weighed against the potential benefit of bypass starters.

2.7 Relay Logic and Ladder Diagrams

Relays may be used to implement control sequences using Boolean logic. Boolean logic is a way of making decisions using two-position switches or states. A switch may be opened or closed, also expressed as On or Off, 1 or 0. This is the way digital computers work; in fact, the name *digital* is derived from the discrete on/off states. The first digital computer was constructed of a huge mass of electromagnetic relays, now long since replaced with semiconductor devices.

Boolean logic or relay logic can be used in simple control sequences. For instance, suppose we wish to turn on a pilot light if either fan A or fan B was on. To sense the fan's status, we could install a differential pressure switch across the fan. The contact would be open if the fan pressure was low and closed if the fan pressure was high. To make the light turn on if either fan was on (if either pressure switch contact were closed), we would wire the two contacts in parallel (*Figure 2-28*, line 1). If we want to turn on the light only if both fan A and fan B were on, the contacts would be wired in series (*Figure 2-28*, line 2).

Using Boolean logic notation, if turning on the light was labeled event C, then these two sequences could be written:

If (A or B) then C   (parallel wiring – line 1)
If (A and B) then C   (series wiring – line 2)

Suppose we wanted the light to turn on when either fan was on, but not if both were on together. Instead of the on/off contacts used above, we could use single-pole double-throw (SPDT) contacts (Section 2.4). *Figure 2-29* shows

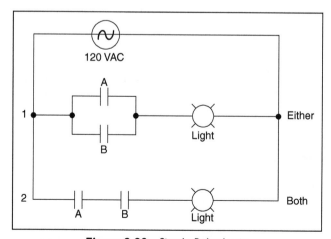

Figure 2-28   Simple Relay Logic

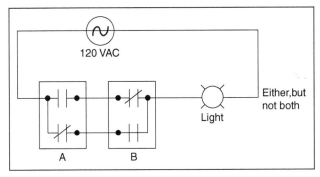

Figure 2-29   Relay Logic with Double-throw Contact

how the double-throw contacts can be used to turn on the light if either but not both fans is on. This would be represented in Boolean format as:

If (A or B) and NOT (A and B) then C

While the fan status contacts in the previous example are from a current sensing relay, the same type of double-throw configuration is also typical of almost any relay or switching device, such as most two-position thermostats and differential pressure switches.

An application similar to the previous example using switches instead of contacts is the three-way switching arrangement common in households, where a light must be switched from either of two locations. We would like the light to change states (go on or off) if either switch is switched. If single throw switches were used in parallel (like *Figure 2-28*, line 1), then either switch could turn the light on, but both would have to be off for the light to go off. If the switches were wired in series (like *Figure 2-28*, line 2), either switch could turn the light off, but both would have to be switched on for the light to go on. The solution is shown in *Figure 2-30* using single-pole double-throw switches.

Ladder diagrams are a type of wiring diagram used to convey control logic such as relay logic and, typically, to show how devices are to be physically wired in the field. *Figures 2-28–2-30* are examples of elementary ladder diagrams. The name comes from the diagram's resemblance to a ladder. The "rungs" of the ladder are wires interconnecting control contacts (such as relay contacts, thermostats, switches, etc.) to loads (relays, starter coils, etc.) across

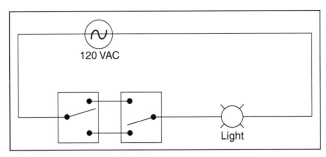

Figure 2-30   Three-way Light Switch

the control power voltage potential represented by the two vertical lines on the left and right.

Figure 2-31 shows many of the symbols used in ladder diagrams to represent commonly used HVAC devices. Many of these devices have already been introduced. One group which has not been introduced is the time delay relay. Often equipment needs to operate in sequence or to have a delay before action is taken.

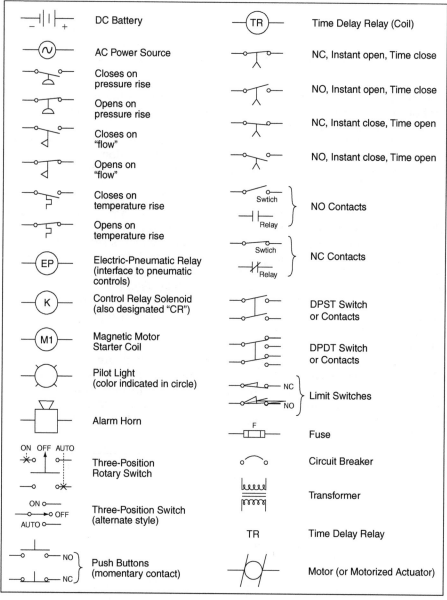

Figure 2-31   Symbols for Electrical Logic Devices

Consider, for example, a simple ventilation unit with a normally closed outside air damper. On startup we may want the damper to be open before the fan starts. We can achieve this with a normally open, time to close, relay supplying the fan starter. When the system is powered on the damper will open while the timer on the fan relay runs and then closes, powering the fan.

Ladder diagrams can have "bugs," inadvertent errors that cause an unexpected result. Most commonly, these bugs are caused by unintentional voltage feedback. A very simple example is shown in *Figure 2-32*. We would like to start two heating fan-coil units from a timeclock contact (TC) provided the spaces they serve are below 70°F, as indicated by the two space thermostats shown. We would also like to start fan-coil 1 if the space temperature dropped below a night low limit thermostat set point (55°F), so we add a parallel connection to the relay that starts fan-coil 1. The bug in this design is that when the night low limit thermostat closes, the contacts of the other thermostats in the space will also be closed because their set points are even higher. This will allow the voltage to feed back through the space thermostats so that if the space served by fan-coil 2 was below 70°F, fan-coil 2 would also start. This was not our intention.

To avoid this problem, another relay could be added, as shown in *Figure 2-33*. As this example demonstrates, care must be taken to avoid feedback when there are multiple, parallel paths to switch a device.

Figure 2-34 is a somewhat more complicated example of a ladder diagram. On more complicated diagrams, such as this, it is common to number the lines for reference. When a relay contact is used in a given line, the line number that the relay coil is on should be referenced (the numbers on the right side of *Figure 2-34* in parentheses) to make it easier for the reader to follow the logic.

The example shown in *Figure 2-34* is for a simple system consisting of a fan-coil, secondary hot water pump and a toilet exhaust fan, all serving a single space. The following is the control logic this diagram represents:

> The fan and pump may be started and stopped manually or automatically using hand-off-auto switches. In the auto position, the fan shall start based on the timeclock or by the night low-limit thermostat (set to 55°F

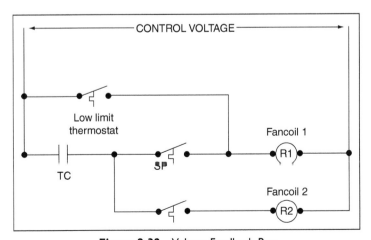

Figure 2-32 Voltage Feedback Bug

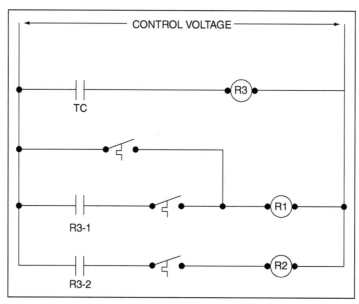

Figure 2-33   Correction for Voltage Feedback

with a 3°F differential). If either the freeze-stat or discharge smoke detector trip, the fan shall stop whether in hand or auto mode. In the auto position, the pump shall start whenever the fan is on, as indicated by an auxiliary contact at the fan starter, and the outdoor air temperature is less than 65°F. When the pump is on, the pneumatic thermostat controlling the hot water control valve (not shown because these are not electrical devices) shall be enabled via an electric-pneumatic (EP) valve. The 120-V toilet exhaust fan shall be started by the timeclock after first opening its discharge damper (powered by a 120-V two-position actuator) and allowing 30 s for the damper to fully open.

A load is energized when the contacts connecting the device to the left-hand power lead to the right-hand lead all close. For instance, looking at *Figure 2-34*, lines 3 and 4, and assuming the H-O-A switch is in the auto position, for the starter coil for the fan-coil (M1) to energize, we would first have to close either contact R1-1 (one of the contacts from relay R1) or the night low-limit thermostat contact (which closes on a fall in temperature below set point, 55°F in this example). That would pass the control voltage through the auto terminal of the H-O-A switch and up to the two safety contacts, the smoke detector and freeze-stat. These contacts are normally closed; meaning they would open in the event of a safety fault (for example, when smoke is detected or when freezing temperatures at the coil inlet are detected). Assuming these safeties are satisfied, the control voltage is passed on to the starter coil. The circuit is complete if all the overload contacts are satisfied, energizing the coil, which would start the fan motor. At the same time, the coil's auxiliary contact is closed. This contact is used to start the hot water pump.

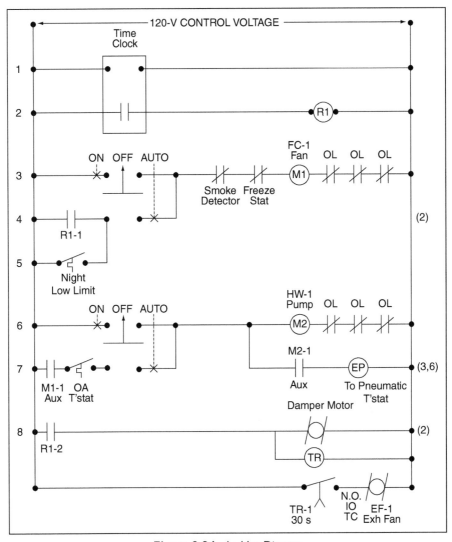

Figure 2-34 Ladder Diagram

While *Figure 2-34* clearly conveys the desired control logic, it actually is a poorly constructed diagram. For instance:

- The diagram implies that there is a single source of control power for all the devices shown. While this is possible, it is not likely. The motor starters probably each have their own control transformers while the toilet exhaust fan is probably served by its own power circuit. To avoid this confusion, it is important that ladder diagrams be consistent with the actual power sources used. In this example, the drawing would have to be broken into several stacked ladders, one for each control voltage source.

- The physical location of devices is not clear. The diagram implies that all the devices are in the same panel. In reality, the timeclock and perhaps the EP valve are in a control panel; the H-O-A switches, overloads, and starter coils are in the individual motor starters; and the thermostats are in other remote locations. This problem is often avoided by labeling devices and terminal numbers to show location.

While the ladder diagram is a convenient means of conveying control logic, it is often a poor representation of the physical layout of the system, as this example illustrates. In most cases, HVAC design engineers will never need to develop a ladder diagram as part of the control system description. Written logic sequences and system schematics, along with coordination details and specifications, are usually sufficient to satisfactorily convey the design intent and division of responsibility to the bidding contractors. Where there may be confusion in some wiring responsibility, a detail such as that shown in *Figure 2-24* should be added to the plans, but a ladder diagram seldom is required.

Ladder diagrams are probably most commonly used with factory installed controls associated with packaged air conditioners and chillers. However, even in this application, they are becoming obsolete as control system designs migrate towards digital controls. With digital controls (as we will learn in Chapters 10–12), much of the control logic is implemented in software, thereby obviating the need for many of the relays and controllers found in electric control systems.

The Next Step

In the next chapter, we will learn about control valves and control dampers, two of the most common controlled devices used in HVAC systems. The proper selection and sizing of control valves and dampers are critical to the performance of the control system. As these devices have very specific operating characteristics we can often choose the control methodology to best suit these characteristics.

Chapter 3

Control Valves and Dampers

Contents of Chapter 3

Introduction

When you have finished this chapter you should have a good conceptual understanding of how controls work. Although we will be providing you with some information about calculations, the issues of sizing and how flows vary under different circumstances may not be clear to you. This is because valve choice and sizing is not a simple process for anyone. In the real world, many designers rely on control specialists (consultants and manufacturers) to choose and size the valves and dampers for their systems. Although the valve choice may be done by someone else, you should understand the options for design and the basic terminology that these specialists will use.

Study Objectives of Chapter 3

Control valves and control dampers are the two primary means to control the flow of water and air in HVAC systems. This chapter explains how these devices work and how they are selected and sized.
   After studying this chapter, you should:

Be familiar with the valve types that are available and which are most appropriate for flow control in various applications.
Understand the operation of mixing valves and diverting valves.

Know how to select control valves for hydronic systems.

Understand the difference between parallel blade and opposed blade dampers and where each is used.

Understand how damper design affects leakage, and where leakage minimization is important.

Understand how mixing dampers can be selected and positioned to encourage mixing.

Know how to size mixing and volume control dampers.

3.1 Two-way Control Valves

The control valve is possibly the most important component of a fluid distribution system because it regulates the flow of fluid to the process under control. In HVAC systems, control valves are primarily used to control the flow of chilled water, hot water, and condenser water, the subject of this section. Control of other fluids including steam, refrigerants, gasses, and oil are similar in many aspects but are not specifically addressed here because they have specific requirements for design including issues of safety and material compatibility.

Styles and Principles of Operation

Control valves may be either two-way (one pipe in and one pipe out) which act as a variable resistance to flow or three-way (two pipes in and one out for mixing valves – one pipe in and two out for diverting valves) as depicted in *Figure 3-1*. Three-way valves may be either mixing (two flow streams are merged into one) or diverting (a single flow stream is broken into two), as shown in the figure. With all three configurations shown, the valves modulate flow through the cooling or heating coil to vary the capacity of the coil.

With the two-way configuration, flow through the circulation system is variable. In the three-way configurations, flow remains relatively constant through the loop which includes the pump and varies in the loop containing the coil. This works well for systems in which the supply of heat, typically a boiler, or supply of cooling, typically a chiller, requires a constant flow. In other systems, it may be that constant flow in the coil is important, perhaps to prevent freezing. In this case the pump can be in the coil loop.

Control valves typically come in three valve styles: globe, butterfly, and ball. The globe-type valve has been the most common for many years, but the characterized ball valves are becoming very popular and are starting to become a significant part of the working marketplace. Below 2-inch size, they have usually sweat (soldered) or screwed connections, while above 2 inch they are typically flanged.

Figure 3-2 shows a typical globe-type two-way single-seated control valve. It consists of a body, a single seat, and a plug. The plug is connected to a stem, which, in turn, is connected to the actuator, also called the actuator or motor. Moving the stem up and down controls the flow. Full shut-off is achieved when the plug is firmly down against the seat.

The body is connected to the piping system in any suitable way (screwed, flanged, welded, soldered, etc.), but it is important that unions or something similar be provided so that the valve can easily be removed for repair or

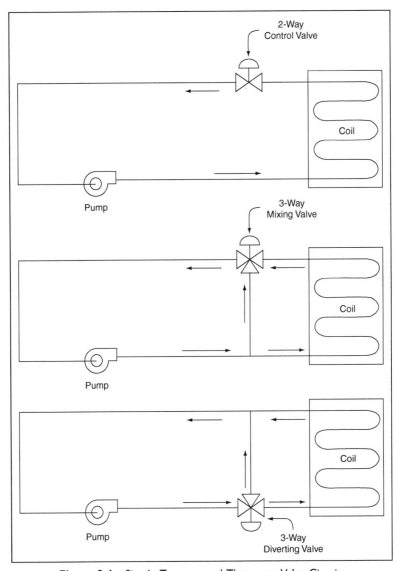

Figure 3-1   Simple Two-way and Three-way Valve Circuits

replacement. Make sure the flow direction is correct with the arrow on the valve body. Service (manual) valves should be provided to isolate individual control valves or piping subsystems.

An actuator that is sprung to lift the valve stem upon power loss combined with the globe valve shown in *Figure 3-2* would produce a normally open valve assembly. The valve is open when power is removed from the actuator.

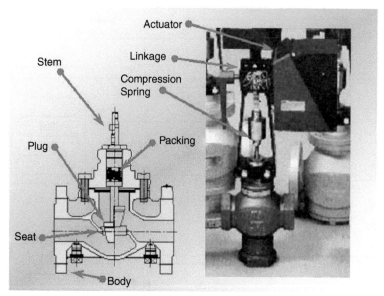

Figure 3-2   Two-way Globe Single-seated Valve (Fluid Flow is Left to Right)

Figure 3-3 shows a globe valve that closes with the stem up. Using this actuator with the valve in *Figure 3-3* would produce a normally closed valve assembly as the valve is closed when power is removed from the actuator. In both cases, the stem must be driven against the flow of fluid to close the valve. Normally open valves are generally desired, when available, as they always fail to the open position, and, if closure is desired, then manual valves can be closed down/off to restrict the flow until repairs can be made.

The figures indicate that flow through the valve must occur in the direction shown by the arrow. All control valves will have an arrow cast into the

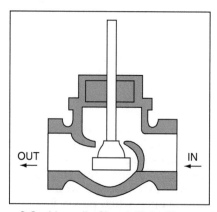

Figure 3-3   Normally Closed Globe Two-way Valve

outside of the body to indicate flow direction. The reason for this is as follows: in any linkage between motor and valve stem there will be some slack, a little free movement of the valve stem. When flow occurs in the correct direction, the velocity pressure of the fluid and the fluid differential pressure across the valve will tend to open the valve. Therefore, the motor must press tightly to close it, taking up any free movement. If flow takes place in the wrong direction, the velocity pressure tends to close the valve (pushing down on top of the plug of the valve in *Figure 3-2*). When the valve throttles toward its closed position, the pressure may be enough to push the plug to the closed position, taking advantage of the free movement or slack in the valve stem. When this happens, flow ceases, then the velocity pressure component disappears, and free movement allows the valve to crack open. Flow begins, the velocity component reappears, and the cycle is repeated indefinitely. Each time the flow stops and starts, the inertial force of the fluid in the pipe causes a shock known as *water hammer*. Besides being noisy and annoying, it can cause failure of the piping system. Therefore, it is important never to install a control valve backward.

Figure 3-4 shows a double-seated valve, also called a balanced valve. As the name implies, it has two plugs and seats arranged so that fluid differential pressure is balanced and the actuator does not have to fight against differential pressure to close the valve, as it does in the single-seated valves shown in the *Figure 3-2*. This reduces the size of the actuator. But the valve inherently cannot provide tight shut-off. This reduces its applicability to HVAC systems, where tight shut-off is usually desired, to minimize energy costs (to prevent leakage and simultaneous heating and cooling).

Modulating globe-type control valves is made with two basic types of plugs: the linear (V-port) plug (see *Figure 3-5*) and equal percentage plug (see *Figure 3-6*). Many manufacturers have variations on these two designs (called modified linear or modified equal percentage), the characteristics of which are usually similar to those described here.

A flat plate plug (see *Figure 3-7*) is sometimes used for two-position, quick-opening duty.

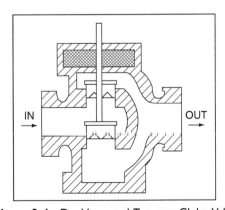

Figure 3-4   Double-seated Two-way Globe Valve

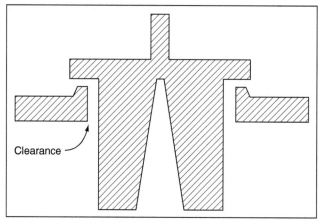

Figure 3-5   Linear (V-Port) Valve Plug

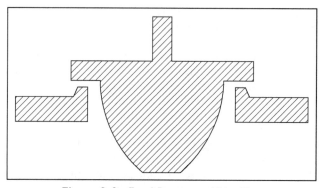

Figure 3-6   Equal Percentage Valve Plug

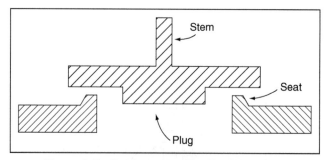

Figure 3-7   Quick-opening (Flat Plate) Valve Plug

The graph in *Figure 3-8* shows the relationship of percent flow to percent plug lift for each plug type, assuming constant pressure drop across the valve. Plug lift is defined as zero with the valve closed, and up to 100% when the valve is opened to the point beyond which no increase in flow occurs. The flat plate plug provides about 60% of full flow when open only 20%. Thus, it is suitable only for two-position control.

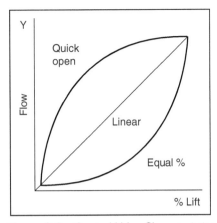

Figure 3-8   Control Valve Characteristics

   Control valve characteristics is a complex study of the what characteristics are needed from the HVAC system and its coil, and how the valve is designed to operate and function. Correctly choosing these characteristics issues can yield a properly combined control valve for its application. A very simple example of this is depicted in *Figure 3-9*.

   As shown in *Figure 3-10*, the linear plug has an essentially linear characteristic while the equal percentage plug is shaped so that the flow increment is an exponential function of the lift increment. This means that when the valve is almost closed, a large percent change of lift is required for a small change of flow.

   As the plug reaches its last tiny increment of closure until it fully shuts, the flow drops off very quickly. This minimum flow rate just before closure is a function of the physical construction of the valve, plug, and seat. The ratio of the minimum rate to maximum rate at the same pressure drop across the valve is called the rangeability or turn-down ratio. For a typical HVAC control valve, this ratio will be about 20:1, which is equivalent to a 5% flow when the valve is barely cracked open. This is usually adequate for HVAC control work. Valves with larger ratios are available but they are more expensive.

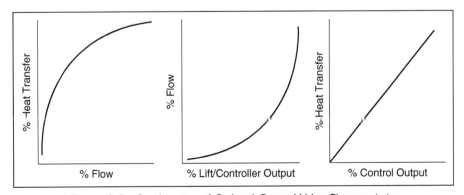

Figure 3-9   Combination of Coil and Control Valve Characteristics

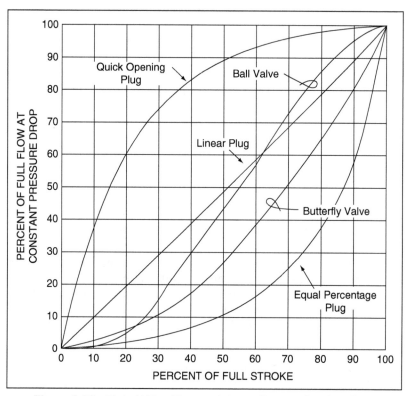

Figure 3-10   Typical Valve Characteristics at Constant Pressure Drop

Figure 3-11 shows a butterfly valve, which is basically a round disk that rotates within the valve body to modulate flow. While not always suitable for modulating duty (as discussed in the next section), butterfly valves can be used for shut-off, balancing, and two-position and three-way duty. The butterfly valve has a characteristic that falls in between the equal percentage and linear plug characteristics, see *Figure 3-10*, while the ball valve has a nearly linear characteristic. Different flow characteristics are desired in different applications.

A ball valve (basically a bored ball which rotates in the valve body) is shown in *Figures 3-12* and *3-13*. Ball valves are primarily used as shut-off and balancing valves on small piping systems (2 inch nominal pipe size and less), but recently they have been adapted for automatic control applications, primarily for small coils such as reheat coils. Ball valves, without an appropriate plug, should not be used in large flow control purposes; typically the resistance, when open, is too low and it lends itself to allowing a much smaller size valve in relation to the pipe, and its control is unstable.

Ball valves with a "characterized plug" can be used in some typical HVAC control applications as depicted in *Figure 3-13*.

The flow characteristics of these ball valve standard and characterized plugs are shown in *Figure 3-14*.

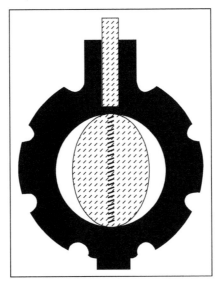

Figure 3-11   Butterfly Valve

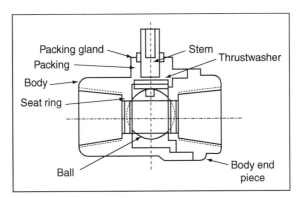

Figure 3-12   Ball Valve Layout

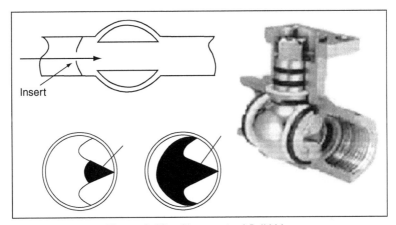

Figure 3-13   Characterized Ball Valve

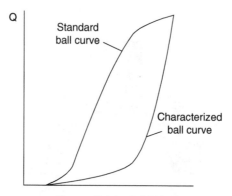

Figure 3-14   Ball Valve

The three types of valves considered – globe, butterfly, and ball – all need driving with an actuator. The globe valve actuator moves the valve stem in and out as shown in *Figure 3-15*. The actuators for ball and butterfly valves must rotate the valve stem with an actuator typically as shown in *Figure 3-16*.

Using two-way valves offers several advantages over three-way valves, including:

The valve is less expensive to buy and install. This is partly offset by the actuators typically costing more because of the higher differential pressure across the valve.

Two-way valves result in variable flow that will reduce pumping energy. This is particularly true when variable speed drives are used on pumps.

Piping heat losses as well as pump energy can be reduced by using the valve to shut-off flow to inactive coils while serving active coils; this is an advantage when a central plant serves many coils operating on different schedules.

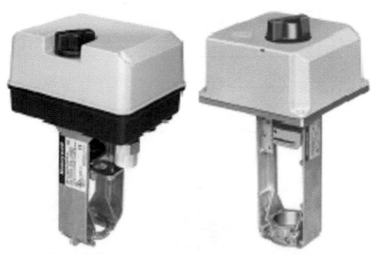

Figure 3-15   Valve Actuators – Move Stem Up and Down

Figure 3-16   Valve Actuator – Rotary (Courtesy Honeywell)

Diversity in load may be taken into account when sizing the pumping and distribution systems, potentially reducing their costs.

The need for system balancing flows is reduced or eliminated in most applications. Because the valves will only use as much chilled or hot water as required by the load, the two-way valve system is self-balancing under normal operating conditions. With three-way valves, flow occurs through the circuit at all times (either through the coil or the bypass), so flow must be balanced to ensure that the required flow is delivered to each coil.

On the other hand, the use of two-way valves can have disadvantages:

Some chillers and boilers cannot handle widely varying flow rates. Using three-way valves in place of two-way valves is one way to resolve this problem. (Two-way valves may still be used at coils, but some other means to maintain flow through the equipment must be included, such as a pressure actuated bypass, VSD, or a primary/secondary pumping system. The reader is referred to the *ASHRAE Handbook – HVAC Systems and Equipment* and other sources for more information on these alternative designs.)

Two-way valves cause differential pressures to increase across control valves, particularly when pumps are uncontrolled (allowed to ride their pump curves (see *Figure 3-31*). This reduces the controllability of the system and may even cause valves to be forced open by the water pressure. Actuators typically are sized larger to handle the much larger pressure close-offs.

Because of the advantages they offer, use of two-way valves is generally recommended, used with the appropriate bypass or VSD design, particularly for large systems where their energy and first-cost advantages are significant. But the system design and valve selection (discussed in the next section) must be able to mitigate these two disadvantages for the system to work successfully.

3.2 Three-way Control Valves

Three-way valves provide for variable flow through the coil while maintaining somewhat constant flow in the system, as shown in *Figure 3-1*.

Mixing and diverting three-way valves are shown in *Figures 3-17*. In a mixing valve, two incoming streams are combined into one outgoing stream. In a diverting valve, the opposite takes place. The exiting port of the mixing valve and the entering port on the diverting valve are called the common port, typically labeled C (for common), or sometimes AB.

In *Figure 3-18*, the bottom port of the mixing valve is shown as normally open to the common port, COM. (open to the common when the stem is up).

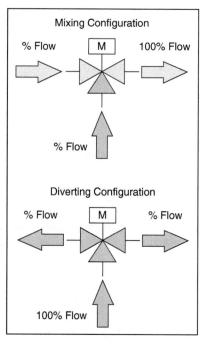

Figure 3-17   Mixing (Left) and Diverting (Right) Valve Configurations

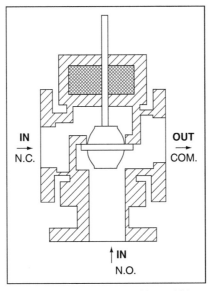

Figure 3-18   Three-way Mixing Valve

This port is typically labeled NO (for normally open), although it is some-times labeled B (bottom port). The other port is normally closed to the com-mon and is typically labeled NC (normally closed), although it is sometimes labeled A or U (upper port). The common outlet is usually labeled COM or OUT. The diverting valve is similarly labeled.

In *Figure 3-19*, the common port of the diverting valve is shown in the same location as that on the mixing valve, on the side.

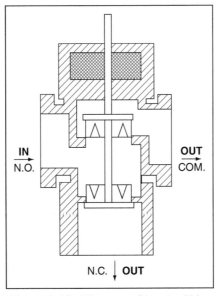

Figure 3-19   Three-way Diverting Valve

With some manufacturers, the valve may be designed so that the common port is the bottom port, with water exiting left and right. Notice that, like two-way valves, the plugs for both mixing and diverting valves are arranged to avoid water hammer (i.e., flow is under the valve seat). Therefore, it is important that the valve be properly piped and tagged with respect to flow direction, and a mixing valve must not be used for diverting service, or vice versa.

Mixing valves are less expensive than diverting valves and thus are more common. In most cases, where three-way valves are desired, they are arranged in the mixing configuration, but occasionally a diverting valve is required.

The more common use of mixing valves over diverting valves is apparently the reason why two-way valves are traditionally placed on the return side of coils (where a mixing valve must go) rather than on the supply side (where a diverting valve would be), as shown in *Figure 3-1*. From a functional perspective, it makes *no difference* on which side of the coil the two-way valve is located. Two-way valves located on the return side of coil piping will maintain pump discharge pressure on hydronic coils to enable positive air venting from the coil return header. Additionally, the fluid passing through the valve on the return side is tempered by the heat loss/gain through the coil.

Figure 3-20 shows two typical three-way mixing valve schematics.

Notice how the valve ports are labeled; it is important that control schematics be labeled in this manner to be sure the valve is piped in the desired configuration so that it will fail to the proper position and respond properly to the control action of the controller. The common port is oriented so that flow always returns to the distribution return. In the example at the top of *Figure 3-20*, the valve is normally

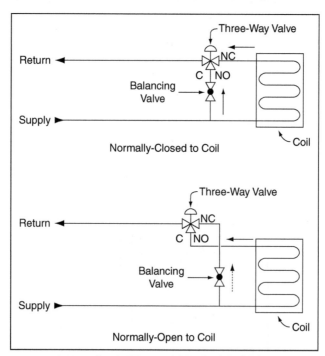

Figure 3-20   Typical Three-way Mixing Valve Arrangements

closed to flow through the coil. If the normally open arrangement was desired, the port labels on the schematic could simply be reversed (the NO label would be shown at the valve return). However, because the normally open port on a real three-way mixing valve is on the bottom, simply relabeling the schematic encourages errors in the field. It is better to rearrange the schematic, as shown on the bottom of *Figure 3-20*, so that the NO port is shown in the proper position.

Notice the balancing valve shown in the coil bypass line of *Figure 3-20*. While not generally a part of the control system (and, as such, it is not typically shown on control schematics), this valve is nevertheless essential for proper operation of the water distribution system unless the coil pressure drop is very low. The valve must be balanced to match the pressure drop of the coil so that when the valve is in the bypass position, the pressure drop will be similar to the path through the coil. Without the valve, a fluid short-circuit occurs and the sup-ply-to-return differential pressure in the system will drop, possibly starving other coils in the system that require a higher differential pressure.

Plugs in three-way valves are available in the same styles as two-way valves, typically linear and equal percentage. However, not all manufacturers make both styles in all sizes, so the designer does not always have flexibility in selection within one manufacturer's line. In some rare instances, valves are built with two different plug styles, allowing the valve to behave in a lin-ear fashion for one port and an equal percentage fashion for the other. Divert-ing valves seem to be available primarily with equal percentage plugs. The selection of plug style is discussed in the next section.

While three-way valves are most commonly used where constant fluid flow is desired, in reality they will not result in constant flow no matter which plug style is selected. As noted above, the balancing valve can be used to ensure that the flow is the same when flow goes 100% through either the coil or the bypass. However, when the valve is in between these two extremes, flow will always increase with a linear plug and, to a lesser extent, with an equal per-centage plug. The reason for this will become apparent when we consider how valves are sized and selected in the next section.

Before selecting and sizing there is one more behavioral characteristic of modulating valves for us to consider. Modulating control valves have an inherent operating characteristic called "rangeability factor." The rangeability factor of a control valve is the ratio of the maximum flow to the minimum controllable flow. This characteristic is measured under laboratory conditions with a constant differential applied to the valve only. A rangeability factor of 10:1 indicates that the *valve alone* can control to a minimum flow of 10%.

The installed ability of the same valve to control to low flows is the "turn-down ratio." In the real system, the pressure across the valve does not stay constant. Typically, as the valve closes the differential pressure across the valve rises. The ratio of the differential pressure drop when the valve is fully open to when it is almost closed is called its "authority." If the pressure were to stay the same the authority would be $P/P = 1$. However, if the pressure quadrupled the authority would be $\frac{1}{4} = 0.25$. The valve turndown ratio is cal-culated by multiplying the inherent rangeability factor times the square root of the valve authority. Hence, a valve that has decent rangeability (say 20:1) but poor authority (say 0.2) will not have good capability to control down to low flows (rangeability $20 \cdot \sqrt{0.2} = 9:1$), and may only be able to provide "'on-off" control over a good part of its flow range.

Many globe style HVAC control valves do not have high rangeability factors; a major manufacturer lists values from 6.5:1 to 25:1 for their range of globe valves from ½ inch to 6 inch. Most characterized ball control valves, however, have very high rangeability factor (usually > 150:1).

3.3 Selecting and Sizing Valves

Control valve selection will depend on the following considerations:

The fluid being controlled. In this section, the discussion is confined to water. If other liquids, such as water solutions (glycol, brine), and special heat transfer fluids are used, corrections must be made for density and viscosity. Information on special fluids is available from the manufacturers. Information on brines is available in the *ASHRAE Handbook – Fundamentals*. Information on steam valve selection and sizing can be found in control manufacturers' catalogs.

Maximum fluid temperature. This will affect the type of packing and the materials used in the body, body trim, and shut-off disk. Manufacturers' literature must be consulted to ensure that the selected valve meets the required duty. This is seldom a consideration in most HVAC applications because the standard materials used by most manufacturers will be satisfactory at 250°F or more, which is higher than the maximum temperature typically found in HVAC heating systems.

Maximum inlet pressure. This will affect the valve body selection. It is usually only a consideration on the lower floors of high-rise buildings above about 20 stories. The taller the building, the higher the inlet pressure can be at lower floors of the building due to static head from the water in the system. Valve bodies are typically classified as ANSI Class 125 or 250. The class is the nominal pressure rating in psig at very high temperatures. At temperatures typical of HVAC classifications, the actual pressure the valve body can withstand will be significantly greater than the nominal rating. Manufacturers' catalogs should be consulted for valve body ratings at actual operating conditions.

Close-off pressure is the maximum differential pressure the valve must close against. This will affect the actuator selection primarily, but it may affect valve style as well. The differential pressure a valve will experience is often difficult to determine, and depends on the details of the system design. This is discussed further below.

Maximum fluid flow rate. The design maximum flow rate, which is typically determined from HVAC load calculations and coil or heat exchanger characteristics.

Valve style: three-way (diverting or mixing) or two-way.

Control mode: modulating or two-position. This will affect the type of valve, and the type of valve plug. Most of the discussion below applies to modulating applications. For two-position duty, standard globe-type control valves with a flat plug can be used, but often they are not the best selection because their globe style bodies have a high pressure drop even when fully open. While pressure drop is desirable for modulating applications (as we will see below), it is neither necessary nor desirable for two-position applications. For piping less than 2 inch nominal pipe size, motorized ball valves are an option that offer lower pressure drop at a similar price. For larger piping (2 inch and larger), motorized butter-fly valves should be used for two-position applications because they have very low-pressure drops and are less expensive than globe-type control valves.

Desired flow characteristic for modulating applications. This will affect the selection of plug type. Not all manufacturers will offer a selection of valves with different plug types, so the designer may not have a choice in all applications. Desired flow characteristics are discussed further below.

Desired pressure drop when the valve is full open for modulating applica-tions. This will determine valve size, which in turn will determine how well the valve will perform from a control standpoint. This is discussed further below.

Turn-down ratio for modulating applications. The standard valves available from manufacturers will provide acceptable turn-down ratios for HVAC applications, so this is seldom a factor. Consult with manufacturers for special applications.

Among the above selection parameters, three parameters (desired flow characteristic, valve sizing for modulating applications using the desired pres-sure drop, and close-off rating required) are critical and discussed further below. We will start with perhaps the most important factor, valve sizing.

Valve Sizing: Valve size in modulating applications affects system behavior, or gain, which affects the ability of the control system to function as desired and expected. It is probably intuitively clear that an oversized valve will not be able to control flow well. As an extreme example, imagine trying to pour a single glass of water using a giant sluice gate at the Boulder Dam. But under-sizing a valve increases the system pressure drop, which leads to higher pump cost and higher energy costs. We must balance these two consid-erations when making valve selections.

The choice of valve size is based on its pressure drop when fully open. The question then is: what pressure drop should be used? Unfortunately, there is no right answer to this question and there are various differing opinions and rules-of-thumb expressed by controls experts and manufacturers. While there is disagreement about the exact value of the desired pressure drop among these authorities, there is general agreement that the control valve pressure drop (whatever it is) must be a substantial fraction of the overall system pres-sure drop in order for stable control to be possible.

One technique for determining the design pressure drop is to consider the subsystem serving the coil. A typical subsystem for a cooling or heating coil in an air handling unit is shown in Figure 3 21. For discussion we will assume that the pressure drop available between the supply and return mains is a constant, although in practice it will vary. A typical cooling coil pressure drop may be about 3 psi (6.5 ft wg). Given the normal installation of the branch pip-ing (with numerous elbows, isolating valves, reducers, and increasers), it is not unreasonable to have a piping loss of 4 psi (8.7 ft wg) between the supply

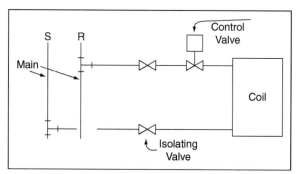

Figure 3-21   Coil Subsystem

and return mains and the coil. Thus, the pressure drop in the subsystem without the control valve is about $3 + 4 = 7$ psi (15.2 ft wg).

From experience and experiment, it has been found that the control valve, in order to be effective in controlling, should have a pressure drop in the range of 50%–100% of the pressure drop through the rest of the coil and piping subsystem at full design flow. Expressed another way, the valve pressure drop should be about 30%–50% of the total subsystem pressure drop. This means that the pressure drop for this valve selection must be from 2 to 3.5 psi. Considering the effect of the higher value on pump horsepower and energy consumption, design engineers usually opt for the lower value (2 psi).

Because determining the pressure drop of each subsystem is difficult and, to a certain extent, arbitrary (there is not always a clear point where a subsystem begins and ends), the logic behind this analysis is often reduced to a more easily used rule-of-thumb, such as simply: the valve pressure drop should be on the order of 2–4 psi. While somewhat arbitrary, this rule-of-thumb tends to work in most typical HVAC applications.

Another rule-of-thumb is to size globe control valves one size smaller than the pipe size. Note that this rule-of-thumb does not work for butterfly and ball valves. Given the variation in pipe sizing techniques among designers, and the variation in flow coefficient (discussed below) among manufacturers, and valve styles, this rule-of-thumb can often result in a poor selection and is not recommended.

With the advent of more sophisticated control algorithms such as PID and fuzzy logic (see Chapter 1), some designers have questioned the need for high valve pressure drops. However, while a well-tuned controller can certainly compensate for some valve over-sizing, there is clearly a point where no control algorithm will help. For instance, getting a single glass of water out of a sluice gate will be impossible no matter how clever the control algorithm may be. Over-sizing will also result in the valve operating near close-off most of the time. This can increase noise from flow turbulence and may accelerate wear on the valve seats. Therefore, relaxing old rules-of-thumb on valve selection is not recommended.

Once the desired pressure drop is determined, the valve can be selected using a rating called the valve flow coefficient, C_v. The valve flow coefficient is defined as the number of gallons per minute of fluid that will flow through the valve at a pressure drop of 1 psi with the valve in its wide-open position, expressed mathematically as:

$$C_v = Q\sqrt{\frac{s}{\Delta P}}$$ (Equation 3-1)

Where Q is the flow rate in gpm, s is the specific gravity of the fluid (the ratio of the density of fluid to that of pure water at 60°F), and ΔP is the pressure drop in psi. Specific gravity for water below about 200°F is nearly equal to 1.0, so this variable need not be considered for most HVAC applications other than those using brines and other freeze-protection solutions. Valve coefficients, which are a function primarily of valve size but also of the design of the valve body and plug, can be found in manufacturer' catalogs.

Using the example above, suppose the coil requires 30 gpm and we have decided to use a 2-psi design pressure drop for the valve. The required flow coefficient would be:

$$C_v = 30\sqrt{\frac{1}{2}}$$ (Equation 3-2)

$$= 21.2$$

Figure 3-22 provides some representative values of C_v for small globe control valves from one manufacturer. Each manufacturer's valves will be somewhat different even if the valves were otherwise similar in style and size. Note that the 21 value falls between the 1¼ inch and 1½ inch sizes. The resulting pressure drop for each valve can be determined by solving Equation 3-1 for ΔP:

$$\Delta P = \left(\frac{Q}{C_v}\right)^2$$ (Equation 3-3)

Representative Values of C_v for Small Control Valves	
Nominal Pipe Size	C_v
1/2	4
3/4	6
1	10
11/4	16
11/2	25
2	40

Figure 3-22   Representative Values of C_v for Small Control Valves

For the 1¼ inch valve, the pressure drop would be 3.5 psi (7.6 ft) while for the 1½ inch valve, the pressure drop would be 1.5 psi (3.3 ft). The latter is marginal (it amounts to less than 30% of the branch pressure drop, excluding the valve itself), so the 1½ inch valve may not provide good control. On the other hand, if the pump capacity was based on the assumption of a 2-psi design pressure drop across the valve, using the smaller valve, while providing better control, could make the pump inadequate. Unfortunately, there are only incremental values of C_v available, so the designer often must make a difficult decision. In general, it is recommended that the selection lean toward the smaller valve rather than the larger because:

Only control valves in the circuit with the highest pressure drop will affect the pump head requirement. All other valves, particularly those closest to the pump where the available pressure differential is usually highest, can be undersized without impacting pump selection or pump energy usage.

Because design loads are seldom encountered, most control valves will operate most of the time over a range from 10% to 50% open. It follows that using a higher design pressure drop through the valve will allow better control most of the time.

In variable flow, two-way valve systems, the differential pressure across the valve can increase, depending on how the pump is controlled. This is particularly true at low flow rates that, as noted above, are the predominant operating condition. Therefore, the valve will appear to be even more oversized as it has to absorb the higher pressure. A smaller valve will improve control under these conditions.

Good control will also affect energy efficiency of the system. For instance, if an oversized valve results in overcooled supply air on a VAV system, the reheat coils at the zone level may have to compensate, increasing heating energy. This may offset any savings gained in pump energy from the reduced valve pressure drop.

An oversized valve on a cooling coil may result in a higher flow and the water leaving the coil at a lower temperature than designed. Returning cold, chilled water (low delta T) can severely reduce chiller plant efficiency.

About the only time where it may be practical to lean toward over-sizing control valves is when supply water temperature is aggressively reset based on system load or load indicators such as outdoor air temperature. This tends to keep flow rates high and minimizes the need for valve throttling. But reset may not be possible under all operating conditions. For instance, on chilled water systems, reset may be limited by the need to maintain dehumidification capability at coils.

For butterfly valves, available flow coefficients are usually very large (pressure drops are very low) and a satisfactory pressure drop may not be possible unless the valve is very small relative to the flow rate. However, for valves this small, the velocity through the valve will usually be above the manufacturer's recommended maximum allowable velocity, above which impingement erosion can negatively affect valve service life. It is for this reason that butterfly valves are not usually the best choice for modulating control duty, where widely varying flow rates are expected and when precise control is required.

When accurate control at low flows is not critical, butterfly valves one or two pipe sizes smaller than the pipe size may be perfectly satisfactory. Two slightly different procedures are adopted in this situation. The first is that valve is sized based on the C_v for the valve when 60% open is commonly used instead of the C_v when fully open as for other valves. The second issue is that when the valve size is smaller than the pipe size the flow is not as smooth. This effectively reduces the C_v of the valve (increases the apparent resistance), and is called the piping geometry factor. For example, a 1½ inch valve might have a C_v of 150 in a run of 2½ inch pipe, 123 in a run of 3 inch pipe, and 80 in a run of 4 inch pipe.

Ball valves also used to have low flow coefficients. This disadvantage has been largely overcome by the use of specially shaped inserts. These inserts restrict the flow through the adjustable ball to give specific characteristics. This use of inserts enables manufacturers to provide, in smaller sizes, a range of off flow coefficients for each valve size.

Note that the discussion above did not distinguish between two-way and three-way valves. From the point of view of the coil, the two are the same; they both result in the same modulation of flow through the coil. Therefore, the same design considerations, and thus the same sizing techniques, apply to both styles. However, pressure differential across the valve and coil in three-way valve, constant flow systems tends to stay fairly constant, while that in two-way valve systems can increase depending on pump controls used (this is discussed in more detail below). Therefore, oversized three-way valves may be slightly more forgiving than oversized two-way valves.

Flow Characteristic Selection

Plug selection depends on the desired flow characteristic, which is a function of:

The heat transfer device being controlled (for example, chilled water coil, hot water coil) and its flow versus capacity characteristic.
The control of fluid supply temperature, which has an impact on the flow versus capacity characteristic of the heat transfer device.
The control of the differential pressure across the valve, which affects the amount of pressure that must be absorbed by the valve.

Figure 3-23 shows typical flow versus capacity curves for heating and cooling coils. For heating coils, the curves can be very nonlinear due to the high temperature of the water compared to the air it is heating. Flow must fall below about 50% before heating capacity falls below 80%. This characteristic becomes even more pronounced when coils are designed for high flow (low temperature drop).

This nonlinear performance can be corrected by resetting the hot water supply temperature, which causes the hot water temperature to draw nearer to the temperature of the air it is heating, thereby effectively reducing the heating capacity of the coil. As can be seen in *Figure 3-23*, the result of hot water temperature set point reset is to linearize the coil's response to flow variation.

Sensible capacity characteristics of chilled water coils tend to be fairly linear inherently due to the closeness of the chilled water and air temperatures.

At low flow rates, the flow in coil tubes becomes low enough that turbulence drops and the flow smoothes out into a flow regime called laminar flow.

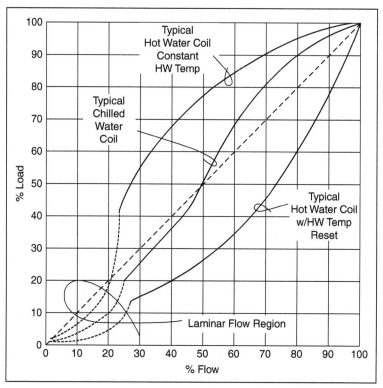

Figure 3-23   Capacity Versus Flow Rate

When flow becomes laminar, the heat transfer coefficient between the water and the inside of the tube suddenly falls, partially insulating the fluid from the air it is heating or cooling. This reduces the capacity of the coil, which tends to linearize the response to flow variation (as shown in *Figure 3-23*), although the results are somewhat unpredictable.

If we were selecting a control valve for a heating coil in a system where hot water temperature would remain fairly constant, control could be improved if the valve flow characteristic (flow versus stroke) was the opposite of the coil's characteristic. In this way, the combination of the two would result in a nearly linear variation of coil capacity with valve stroke. As we saw in *Figure 3-9*, the equal percentage plug would best offset the heating coil's characteristic and is thus recommended for this duty.

If we were controlling a hot water coil in a system with hot water reset, or if we were controlling a cooling coil, the coil's characteristic curve would be nearly linear, as shown in *Figure 3-9*. It may then be logical to use a linear plug to achieve best control. That would be the case with a constant flow (three-way valve) system. But for most two-way valve systems, the differential pressure across the valve will increase as the valve begins to reduce flow.

There are two main reasons for this increase in pressure across the valve:

1. Reduction in pressure drops in other components due to reduction in flow.
2. Increase in available system head at reduced flow.

Let us consider the reduction in pressure drops in other components due to reduction in flow. In an earlier example, the pressure drops were given as coil 3 psi, pipework with fittings 4 psi, and control valve 3.5 psi, all at full flow. The flow for this example is 30 gpm and our system provides an almost constant head of 10.5 psi.

The pressure drop across a fixed component is typically, proportional to the square of the flow. Pressure drop $P = K \text{ gpm}^2$

For our coil, $P = 3$ psi at 100 gpm. $3 = K \times 100^2$, so $K = 0.0003$

At 50% flow $P = 0.0003 \times 50^2 = 0.75$ psi, one-quarter the pressure drop.

In general, if the flow is reduced to 50% the pressure drop will drop to 25%

Now our circuit has a constant head, so as the flow reduces and the pressure drop across the coil and pipework reduces the head across the valve increases. This is shown in *Table 3-1* where the valve pressure drop increases from 8 psi to 16 psi as the valve moves to 50% flow in each branch. You may have noticed that the flow at 50% and 50% is given as 132 gpm, not 50 + 50 =100 gpm. The reason for this is that the linear characteristic of the valve occurs with constant head across the valve. Since the head across the valve has increased the characteristic is distorted to let more water through.

The second issue was the increase in available system head at reduced flow. Consider a very simple circuit including piped coil, two-way control valve, and a pump. The pump will have a flow performance curve with maximum pressure at no flow and the pressure gradually dropping as the flow increases. Our piped coil has a characteristic flow/head curve which obeys the relationship $P = K \text{ gpm}^2$. Initially, at design flow, the valve has a head loss equal to the piped coil. The situation is as shown in *Figure 3-24*.

With the valve fully open the flow will be at point A. The pump head matches the head across the valve plus the head across the piped coil. Now, as the valve closes, the flow reduces and the pressure drop through the piped coil reduces. At the same time the pump flow is reduced, and its head increases. The result is flow at point B and the pressure drop across the valve has approximately doubled.

One way of reducing this effect is to maintain a constant head from the pump by reducing the pump speed. Suppose this did, in fact, maintain the head at the original full at point A flow head, as shown by the broken line. The valve head, pressure differential, is still increased as the flow in the piped coil reduces with reducing flow.

The increase in pressure differential tends to distort the plug characteristic curve because more flow will go through the valve at the same valve position compared to the flow at a fixed differential. This is called authority distortion.

Table 3-1 Pressure/Flow Variations in Control Circuit

Item	Pressure Drops in Coil Circuit (psi)		
	100% Through Coil	50% Through Coil	0% Through Coil
Pipe and fittings	4	1.0	0
Coil	3	0.75	0
Linear control valve	3.5	8.75	10.5
System head (constant)	10.5	10.5	10.5
Total flow coil and bypass gpm	30	~40	30

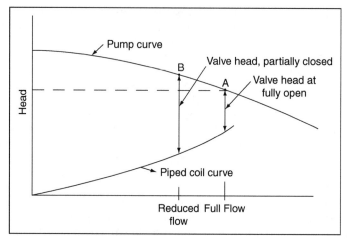

Figure 3-24   Change in Head Across a Control Valve as it Closes

This distortion can be seen in *Figure 3-25* for a linear plug and *Figure 3-26* for an equal percentage characteristic. The variable A (*authority*) in the figures is defined as the ratio of the differential pressure across the valve when the valve is full open to the maximum differential that may occur when the valve throttles to near fully closed. When this ratio is 1.0, the characteristic curve matches that

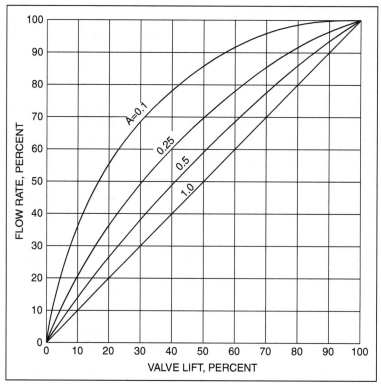

Figure 3-25   Authority Distortion of Linear Flow Characteristics

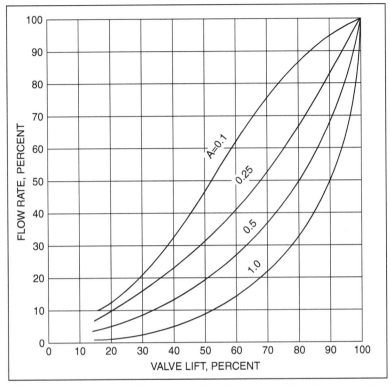

Figure 3-26 Authority Distortion of Equal Percentage Flow Characteristics

shown in *Figure 3-9*. As the ratio reduces (as the maximum pressure increases), the curves become distorted. The linear plug begins to behave more like a flat plug while the equal percentage plug begins to behave more like a linear plug.

For this reason, it is recommended that two-way valves use an equal percentage plug regardless of the type of heat exchanger being controlled. For hydronic applications, the linear plug should only be used for three-way valves controlling fairly linear heat exchange devices such as cooling coils.

Close-off Pressure

As noted above, the close-off pressure is the maximum differential pressure seen by the valve as it closes. Because of the way the valve is configured (see *Figures 3-2* and *3-3*), differential pressure across the valve will tend to push the valve open. The valve/actuator combination must have a close-off rating that is greater than the maximum differential pressure expected. Many valves will have two close-off ratings, one for two-position duty and another for modulating duty that is sometimes called the dynamic close-off rating. The dynamic rating (which is always lower than the two-position rating) is the maximum differential pressure allowed for smooth modulation of the valve, particularly near shut-off. Above this differential pressure, the design turn-down ratio will not be achieved. This is the rating that should be used when selecting a valve for modulating applications.

In practice, it is sometimes difficult to determine the close-off pressure because, as flows in systems change, pressure losses in piping and coils change, and the pump head may also change as the pump rides its curve.

For nominally constant flow three-way valve systems, the differential pressure will usually peak when the flow is fully flowing through coil. As noted in the previous section, under mixing conditions for linear characteristic valves, flow will actually increase through the circuit, which increases pressure drops in the branch lines serving the coil and may cause the pump to ride out on its curve. Both of these effects will tend to reduce differential pressure. Therefore, the close-off pressure for three-way valves is simply the design differential pressure across the coil/valve assembly.

For two-way valve systems, the differential pressure will be at least as high as the design differential pressure across the assembly, but it can be much higher depending on how the pump is controlled. For systems that have pump control using variable speed drives, pump staging, or "choke" valves (valves designed and controlled to restrict flow under specific circumstances), the maximum differential pressure can be determined from the location of the differential pressure sensor used in the pump control loop relative to the valve in question, and the design differential pressure set point. For valves closer to the pump than the sensor location, the differential will be higher than the set point. A conservative approach would be to assume that the flow in the distribution system is at design conditions except the valve in question is closed. The differential pressure can then be calculated by determining the pressure losses backward from the sensor location to the valve.

For variable flow systems without pump controls, the maximum differential pressure will be even higher because the pump will ride up its curve at low flows (point B), increasing available pump head. This can be seen again in *Figure 3-27* for a theoretical one-valve/one-pump system where the maximum differential pressure is simply the shut-off head of the pump.

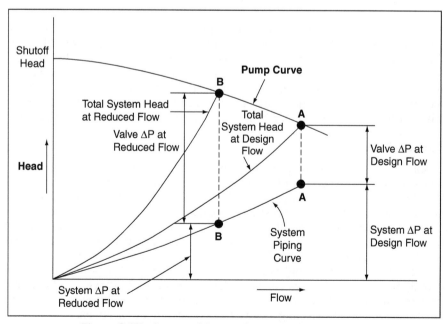

Figure 3-27   Pump and System Curve with Valve Control

The pump shut-off head (the pump head at zero flow) can be obtained from the manufacturer's pump curves. For real systems with multiple valves, the maximum differential pressure will probably be less than the pump shut-off head because the system should never be designed to cause the pump to operate at zero flow for long (it will heat up and the seals will be damaged). However, because the actual differential pressure is difficult to determine, it is practical and reasonable to require all valve close-off ratings to be greater than the pump shut-off head. Often a safety factor (such as 25%) is added just to be sure that valves can close-off tightly.

If a single-seated valve and operator combination cannot be found that meets the desired close-off rating (unusual), a double-seated valve may have to be used. As noted in the previous section, double-seated valves, particularly larger valves (2 inch and larger), will have higher close-off ratings than single-seated valves because the differential pressure across them is balanced. However, they will not provide tight shut-off, which can be a significant disadvantage in most HVAC applications.

3.4 Control Dampers

Dampers are to air as valves are to water: a means of controlling airflow. Many of the design and selection principles are the same for both. Like valves, dampers must be carefully selected and sized to ensure stable and accurate control.

Styles and Principles of Operation

Dampers are used to direct or modulate flow. They may be round, rectangular, or even oval, to suit the duct. Round or oval dampers are almost invariably single blade with a central axle. Rectangular dampers are usually made in sections, with individual blades 6–8 inches wide linked together to move in unison.

Dampers for HVAC work are normally made of galvanized steel or extruded aluminum. Aluminum is preferred on outdoor air intake dampers due to its resistance to oxidation. Other materials are available, for example stainless steel, for use in corrosive atmospheres such as in industrial facilities. Frames and blades must be heavy enough to operate without warping or twisting. Shaft bearings should be permanently lubricated bronze, stainless steel or PTFE, polytetrafluoroethylene – which Dupont have trademarked as Teflon® –to minimize friction.

Blades come in three common shapes: a flat, one-piece (single metal sheet) blade; a single-skin blade with a triple-v-groove shape; and a double-skin airfoil-shaped blade. The triple V and airfoil blades are shown in *Figure 3-28* with external linkages. The face area of the damper is F1 by F2 and frame depth is D. Blade width is typically about the same as the frame depth. The flat blade is typically used only for single-blade dampers in round and oval ducts. The latter two blade types are used in rectangular dampers. The more expensive air foil shape reduces pressure drop and noise caused by turbulence as air passes over the blades. The triple-v-groove blade is typically rated only up to 2000 fpm, and possible noise problems must be considered above about 1500–1700 fpm.

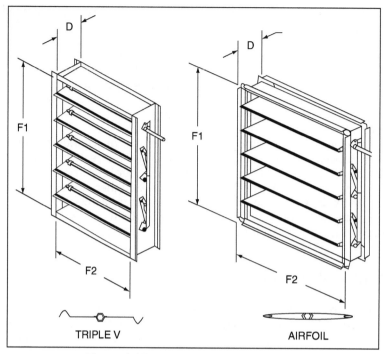

Figure 3-28   Triple V and Airfoil Dampers

Blades are made to overlap and interlock for tight closure. To reduce leakage, a compressible sealing strip may be attached to the blade edges. The material used varies from inexpensive foam rubber to longer-lasting silicone rubber or extruded vinyl. The seals can significantly modify the damper performance particularly as the damper nears fully open and fully closed. Jambs (where the blades align on each side with the frame) may also be sealed to reduce leakage, typically by using a compressible metal or vinyl gasket. Leakage through a standard damper may be as high as 50 cfm per square foot at 1 inch pressure. Low leakage dampers (which usually use air-foil blades) leak as little as 10 cfm per square foot at 4 inch pressure. Shut-off dampers that are normally used in HVAC systems are low leakage type, which usually leak around 2 cfm per square foot at 1 inch wg. Leakage of air through dampers causes false control readings resulting in poor control of the controlled variables. Leakage also causes energy waste and ultimately money. ANSI/ASHRAE/IESNA Standard 90.1-2004, Energy Standard for Buildings Except Low-rise Residential Buildings prescribes a minimum leakage rate and requires testing in accordance with AMCA 500. Table 3-2, which is the information from Table 6.4.3.3.2 in Standard 90.1, gives these rates. A leakage of 4 cfm/ft^2 is an ultra-low leak damper and a 10 cfm/ft^2 is a low leakage damper. The ultra-low leakage requirements are for places with very high cooling loads.

Table 3-2 Maximum Damper Leakage (ANSI/ASHRAE/IESNA Standard 90.1-2004)

	Maximum Damper Leakage at 1.0 in wg cfm/ft² of Damper Area	
Climate	Motorized	Nonmotorized
1, 2, 6, 7, 8	4	Not allowed
All others	10	20(a)

Note: (a) Dampers smaller than 24 inches, in either dimension, may have leakage of 40 cfm/ft²

Linkages are required on multi-blade dampers to make the blades open and close in unison. On marginally less expensive dampers, the linkage is attached directly to the blades, exposed to the air stream as shown in *Figure 3-28*. On marginally more expensive dampers, linkages are concealed in the frame and typically involve rotating the shaft through the blade. Keeping the linkage out of the air stream reduces pressure drop and minimizes the effects of corrosion. It also provides a higher strength interconnection between blades, which provides for a better seal when dampers are closed, particularly after the damper has aged and blades begin to stick. Exposed linkages under these conditions tend to cause the blades to bend and warp so that they will not close tightly. Dampers should be stroked open and closed as a periodic maintenance function. Actuators for dampers should be oversized as, with age, the dampers can become harder to open and close. Periodic maintenance of dampers and their linkages is very important to continuing proper operation. Damper systems that do not move frequently should have a periodic testing and verification of movement at least semi-annually.

For multiple blade dampers, blade movement may be parallel or opposed, as shown in *Figure 3-29*. With parallel blade operation, the blades rotate in the same direction so that they stay parallel to each other throughout the stroke from fully open to fully closed. With opposed blade dampers, the direction of rotation alternates every other blade. These two arrangements have different operating characteristics, a factor in damper selection as discussed in the next section. Operating characteristics of single blade dampers fall somewhere between those of multiple parallel and opposed blade dampers.

Damper motors (also called operators or actuators) must have sufficient power to move the dampers from the closed position, where maximum friction occurs, to fully open. Linkages are sometimes used to couple the dampers to the actuators as shown in *Figure 3-30*. When used in modulating applications, the damper motor must be able to modulate the dampers smoothly through small increments. Low leakage dampers typically require larger operators to overcome the friction of the jamb seals, and parallel blade dampers typically require larger operators than opposed blade dampers. Manufacturers rate operators in inch-pounds of torque. They also provide the torque data for their dampers in terms of inch-pounds of torque per square foot of damper area. For example, the requirement could be for 8 inch-lbs per square foot with a minimum torque of 20 inch-lbs. The minimum torque is to deal with the load when getting the damper moving against the friction of the seals.

As a general rule, dampers and their actuators should be set up for normally open operation. If power is lost, the damper will open and allow its flow to pass

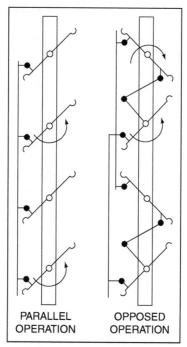

PARALLEL
OPERATION

OPPOSED
OPERATION

Figure 3-29   Typical Multi-blade Dampers (Linkage in the Airstream)

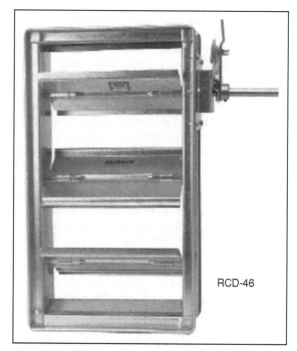

RCD-46

Figure 3-30   Damper Linkages

until repairs can be made to its automatic operation. The exception to this would be dampers protecting coils from freezing, and dampers guarding dangerous, hazardous waste, or isolation systems.

3.5 Selecting and Sizing Dampers

There are three basic damper applications:

- Two-position duty, such as shut-off of outdoor air intakes, fan isolation, etc.
- Capacity control duty, such as air balancing and VAV discharge dampers.
- Mixing duty, such as economizer dampers.

Two-position Duty

Dampers are commonly placed in outdoor air intakes, fan intakes and discharges, and other positions where they are used to prevent air flow from occurring when the fan system is shut off. These dampers have two-positions: they are open when the system is on and closed when the system is off. For instance, a damper placed in an outdoor air intake might be interlocked to the supply fan to close when the fan is off to minimize air infiltration caused by wind or stack effect, thereby protecting coils from freezing in cold weather and reducing energy costs.

Dampers in exhaust fan inlets or discharges are another common shut-off application. These dampers may be motorized or they may be gravity dampers (also called backdraft or barometric dampers), meaning they work without actuators. Gravity dampers seldom offer as tight a seal as motorized dampers. They also only work to prevent air flow in one direction but not the other, and thus may not always achieve the desired result. For instance, on an exhaust fan, a backdraft damper may prevent outdoor air from coming into the building backward through the exhaust air discharge point when the fan is off, but it will not prevent air from going out that discharge point in the normal direction of flow. In a tall building, during cold weather, stack effect can cause enough air pressure to push the gravity damper open, allowing building air to flow out the exhaust system, with corresponding makeup air drawn into the building at lower floors. In these applications, a motorized damper is therefore recommended.

In specific situations, gravity dampers are preferred to motorized dampers. For instance, two fans operating in parallel (see *Figure 3-31*) generally are fitted with dampers to isolate the fans so that one may operate without the other.

This will allow lead/lag operation in variable air volume (VAV) applications (staging of fans so that only one operates at low loads), and provide redundancy by allowing one fan to operate in case the other fails. Without some type of shut-off damper, the operating fan would cause air to flow backward through the inoperative fan (analogous to the need for check valves in parallel pumping systems).

In this application, if a motorized damper is used, there is a question as to when the damper should open. If the damper is opened before the fan is turned on, then while the damper is opening, air will backflow from the other fan, which may already be operating. This will cause the fan wheel to spin

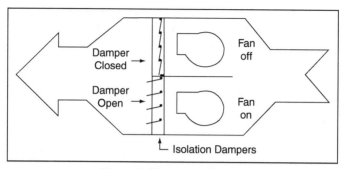

Figure 3-31   Fans in Parallel

backwards, usually causing a motor overload once the fan is turned on and tries to reverse the rotation. If the damper is opened after the fan has turned on, excessive pressure will build up in the fan plenum, possibly resulting in damage. Therefore, a gravity backdraft damper is preferred in this application. It will open automatically when the fan is on and be able to build up sufficient pressure against the backpressure provided by the other fan. Because of the generally high velocity and turbulence near the fan discharge, the damper should be a heavy-duty damper (one that does not "flap in the breeze").

For two-position, and shut-off applications, it generally makes no difference whether the damper uses parallel blade or opposed blade operation because the two look the same when the damper is fully open or fully closed. Parallel blade dampers are commonly used because they are often less expensive. However, in some cases, while the damper itself may be somewhat less expensive, the damper assembly as a whole may not be because parallel dampers often require a larger operator than a similarly sized opposed blade damper. In most two-position applications, the HVAC system designer should therefore allow the vendor to select the least expensive type.

Damper sizing in two-position applications is not critical because the damper is not used for modulating control purposes. The larger the damper, the larger the leakage and the higher the damper cost, but the lower the pressure drop and associated energy costs. Shut-off dampers are typically sized to fit the duct or opening in which they are mounted.

Capacity Control Duty

Probably, the most common application of dampers is to balance flow in air systems. Except under very unique conditions, it is not possible to design duct systems to deliver the desired air quantities to each terminal through the sizing of ductwork alone. Balancing dampers (also called volume dampers) are installed to adjust the pressure drop of duct branches and sections so that design air quantities are provided at each diffuser or grille. Volume dampers are static dampers, meaning they do not have actuators and are meant to be adjusted one time when the system is balanced. Selection and sizing of volume dampers are not critical because they are meant to be trimming devices; they generally need only throttle flow over a small range near fully open. Typically, volume dampers are single-blade, single-skin dampers sized simply to be equal to the size of the duct within which they are placed. Since these dampers are often very economically made they may generate some air noise and are best mounted in the duct as far as possible from any outlets. The length of duct from damper to outlet reduces the noise effect in the occupied space. No special leakage reducing gaskets are used because they are never expected to be fully closed.

Dampers are also sometimes used to control the capacity of fans in VAV fan systems. Two styles of dampers are used: inlet vanes and discharge dampers. Inlet vanes are a special type of damper mounted in the fan's inlet (see *Figure 3-32*). Fitting into the round fan inlet bell opening, the damper blades are pie-shaped and all rotate in the same direction. The effect of the damper is to impart a pre-rotational spin to the entering air. The damper must be installed so that this spin is in the same direction as the fan wheel rotation. With the air moving in the same direction as the wheel, the wheel does less work on the air and thus the fan is effectively unloaded, reducing air volume and static pressure as well as fan energy. Typically, in practice, inlet vanes can vary the flow and pressure from approximately 30–100% of it maximum rating.

Discharge dampers are analogous to "choke valves" on pumps. They simply absorb the excess fan pressure so that the VAV boxes do not have to, thereby allowing the VAV boxes to control air flow to zones in a more stable manner. Discharge dampers do not change the performance of the fan because the fan cannot tell the difference between the pressure drop created by the discharge damper or that created by the VAV boxes. Because of their poor energy performance and potential fan noise problems, caused by fan throttling, discharge dampers are almost never used in modern VAV systems.

For dampers being used for capacity control (throttling), such as volume dampers and discharge dampers, opposed blade operation is preferred to parallel blade operation. The primary reason is that they were believed to exhibit a more linear flow characteristic; the flow through them being nearly a linear function of stroke from fully open to fully closed in throttling applications. The classic diagrams can be seen in *Figures 3-33* and *3-34*. The parameter α in these figures is the ratio of the total system pressure drop to the pressure drop across the damper when it is wide open.

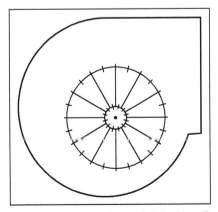

Figure 3-32 Centrifugal Fan with Inlet Vane Damper

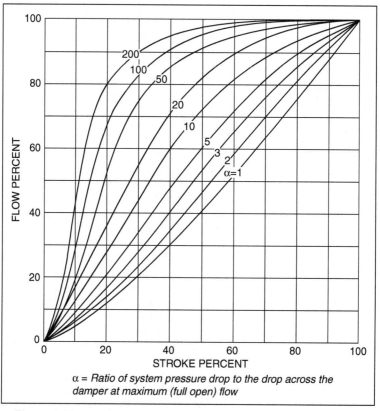

Figure 3-33   Installed Characteristic Curves of Parallel Blade Dampers

Note that this is the same situation that we had with the control valve authority. The only difference (designed to confuse you!) is that:

$$\alpha = \frac{\text{system resistance}}{\text{open damper resistance}}$$

And the valve authority was the inverse

$$\text{Valve authority} = \frac{\text{open valve resistance}}{\text{pipe loop resistance}}$$

These figures are, in fact, a gross and often incorrect picture of damper performance. The movement upwards as authority increases is conceptually correct but the scale of the change may be much smaller than shown in *Figures 3-33 and 3-34*. Damper performance depends on several factors:

1. Manufacturer: due to variations in design including material, linkages, and blade seals. *Figure 3-35* shows the test performance of two triple V parallel blade dampers from different manufacturers. The curves are for the same anti-parallel arrangement of louver and damper in an inlet and a relief arrangement.

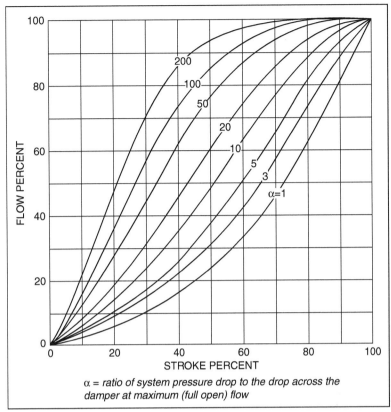

α = ratio of system pressure drop to the drop across the
damper at maximum (full open) flow

Figure 3-34   Installed Characteristic Curves of Opposed Blade Dampers

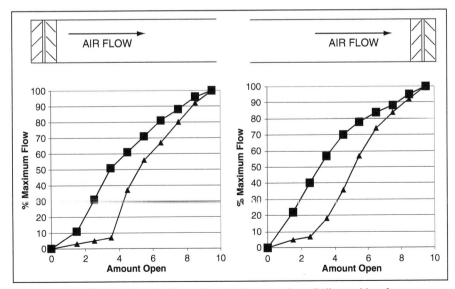

Figure 3-35   Two Parallel Blade Triple V Dampers from Different Manufacturers

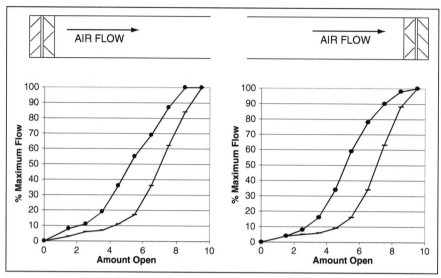

Figure 3-36   Two Opposed Blade Triple V Dampers from Different Manufacturers

The effect for opposed blade dampers in a similar mounting position is shown in *Figure 3-36*. Again two manufacturer's dampers were tested to show the significant variation between specific manufacturers. The tested performance shown in *Figures 3-35* and *3-36* indicates the importance of using manufacturers test data rather than relying on the attractively easy oversimplification shown in *Figures 3-33* and *3-34*.

2. Damper relative size: how large the damper is compared to the duct or wall opening affects the flow. A simple example is the situation where the damper is the same size as the duct so the airflow is relatively straight into the damper. In contrast, a small damper in a large wall will have air coming from all directions into the damper creating a different flow characteristic.
3. Damper situation: any changes in duct direction or devices that can change the air stream mounted before, or after, the damper can make a significant difference to performance. *Figure 3-37* shows the same opposed blade damper performance.

In capacity control (throttling) applications, the pressure drop across the damper will increase as the damper closes because, as air volume reduces, frictional losses in the other parts of the duct system will quickly fall as the square of the air-flow rate, and fan pressure will typically increase as volume falls depending on the shape of the fan curve. (This is analogous to a two-way valve in a hydronic system.) In general, but subject to the specific manufacturer's design, parallel blade dampers will operate in a linear manner in this application only when the pressure drop across the damper is at least 20% of the total system pressure drop. On the other hand, opposed blade dampers will be linear at a much lower pressure drop of around 5% of the total system pressure drop.

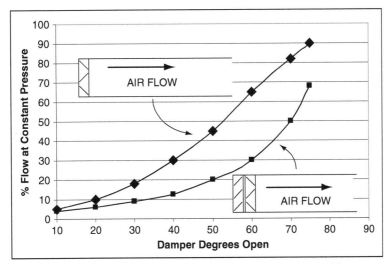

Figure 3-37   Effect of Inlet Louver on an Opposed Blade Damper Characteristic

Another reason why opposed blade dampers are preferred in this application is that they produce less turbulence downstream from the damper. This can be seen in *Figure 3-38*.

The parallel blade dampers deflect the air stream, causing a high degree of turbulence downstream. If a duct fitting such as an elbow was located directly downstream, the pressure drop through it would be much higher than expected due to the asymmetric entering velocity profile. If a diffuser were located directly downstream, the outlet throw pattern and noise generation would be adversely affected. This is why air outlet manufacturers provide opposed blade dampers where dampers must be located in the neck of diffusers.

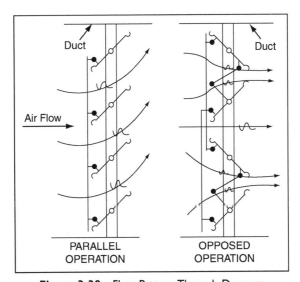

Figure 3-38   Flow Pattern Through Dampers

Damper sizing considerations for a throttling application (such as fan discharge dampers in a VAV system) are similar to those for sizing control valves in hydronic systems; controllability must be balanced with increased pressure drop and associated fan and energy costs. To achieve something approaching a linear response, the opposed blade damper must have a wide open loss of 5%–10% of the system loss. Damper manufacturers' pressure loss data must be consulted for exact damper loss characteristics.

Mixing Duty

Dampers are often used to mix two air streams, with the most common application being the outdoor air economizer system. Economizers work by mixing outdoor air and return air to control supply air temperature instead of or in conjunction with mechanical cooling systems.

In the past, many engineers have mistakenly assumed that because opposed blade dampers are preferred for throttling duty, they must also be preferred for mixing duty. This is usually incorrect. In most cases, it is best to use parallel blade dampers for mixing applications for the same reasons they are not desirable for throttling duty: they tend to deflect the air streams and in this situation they provide good control.

The fact that parallel blade dampers deflect the air flow is a disadvantage in throttling systems, but it is a positive feature in mixing applications. As shown in *Figure 3-38b*, the parallel blades may be oriented so that they deflect the two air streams into each other and encourage mixing. With opposed blade dampers, the two air streams will tend not to mix, as shown in *Figure 3-39a*. The result is stratification, which means that there are two different temperatures in the duct: one close to outdoor air temperature and one close to return air temperature. Stratification can continue through the duct system for long distances, even through a centrifugal or axial fan. Stratification makes measurement of the air temperature more difficult and less accurate, it can reduce system ventilation effectiveness by allowing some areas to get less outdoor air than others, and in cold climates it can result in the freezing of coils. In addition to using parallel blade dampers, mixing can be enhanced by physically orienting the outdoor air and return air entries into the mixing plenum to direct the air streams into each other. Various successful arrangements are shown in *Figures 3-39(c)–(f)*.

When the opposed blade damper is at 50% stroke, the actual area of air path opening between the blades is much less than 50%. This characteristic is what makes an opposed blade damper desirable in a throttling application where the pressure across the damper increases as the damper closes. But it makes the damper very undesirable in a mixing application. This can be seen in *Figure 3-40*, which shows the pressure drop across a typical mixing box assembly with both opposed blade and parallel blade dampers. At 50% stroke, the opposed blade dampers will cause the pressure drop to rise to three times that when the dampers are fully open to the outdoor air or to the return air. Other experiments have shown this increase to be as large as 700% (Alley, 1988). Parallel blade dampers, on the other hand, tend to cause the pressure drop to fall slightly in the mid-position.

In constant volume applications, we would like the pressure drop across the fan to remain constant. If the pressure drop increases, the flow rate to the

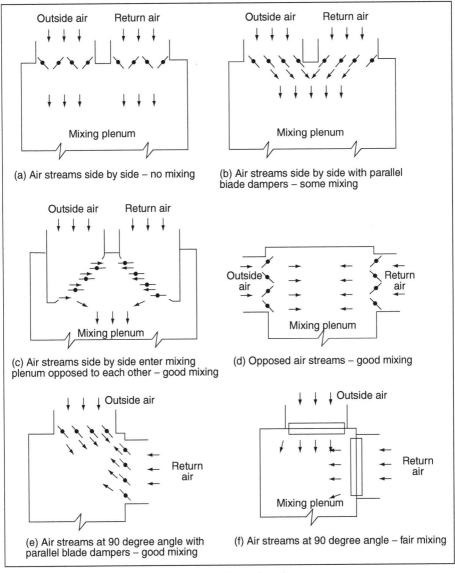

Figure 3-39 Various Mixing Box Arrangements

space will fall, possibly causing comfort problems. If the pressure drop falls, the flow rate will generally increase depending on the shape of the fan curve, thereby increasing fan energy and possibly increasing noise from ductwork and air outlets. Clearly, from *Figure 3-40*, using opposed blade dampers will have a negative impact on flow and should not be used. Parallel blade dampers also will affect flow, but only in a minor way. To achieve the ideal, a constant pressure drop across the fan (the performance curve labeled C in *Figure 3-40*) must be achieved. It causes pressure drop across the mixing assembly to slightly increase in the mid-position. This is to compensate for

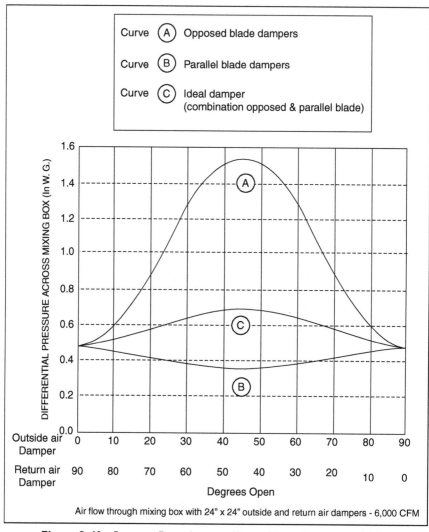

Figure 3-40 Pressure Drop Across a Typical Mixing Box (Avery, 1986)

reducing pressure drop through the outdoor air intake louver, which varies as the square of the outdoor air intake volume. This ideal performance character-istic could be achieved by using a mixture of opposed blade and parallel blade dampers. However, doing so adds complication and may cause confusion in the factory and field. Thus, it is not recommended unless precise flow control is required. Using parallel blade dampers only will cause a slight increase in fan volume at mid-stroke, but this is generally acceptable in most applications.

For variable air volume applications, the fact that the pressure drop across the mixing assembly drops with parallel blade dampers, can be a source of energy savings, because the fans generally have volume control devices such as variable speed drives or inlet vanes. Therefore, in general, where practical, parallel blade dampers are preferred for VAV applications.

Sizing considerations for mixing applications are very different from throttling applications. This is because the pressure drop across the assembly does not increase due to changing system pressure drops (as it does in throttling applications), and because the desired result is the mixture of two air streams, not the reduction of one air stream. Mixing dampers are analogous in performance to three-way valves. With three-way mixing valves controlling a coil, the desired result is maintaining a required temperature off a coil by modulating the flow through two inlets from the coil and bypass. With an economizer mixing assembly, the desired result is a specific temperature of mixed air achieved by modulating the two airstreams. The outlet flow rate is, ideally, constant from both the three-way valve and mixing dampers. This means that over-sizing dampers for mixing applications is less of a problem than it is for throttling applications.

The size of the outdoor air damper is a function not only of the maximum flow rate through it but of the differential in temperature between the two air streams. The closer the two air stream temperatures are to each other, the less the impact sizing will have on controllability. In very cold climates, where a small amount of outdoor air may have to be metered into the return air stream to achieve the desired supply air temperature, the outdoor air damper should be smaller than it need be in mild climates. Similarly, in hot humid climates, where good control of minimum outside air is required, a smaller damper is often chosen. The rule-of-thumb for sizing outdoor air dampers is to achieve a face velocity of around 1000–2000 fpm, the higher value to be used in colder climates. It is also not uncommon, due to space constraints, to make the damper the same size as the intake louver, typically around 400–500 fpm. This very large damper will only cause control problems if it is used to control a small amount of outdoor air, such as might be required for minimum ventilation (on the order of 15%–25% for typical applications). This problem can be avoided by using a separate two-position minimum outdoor air damper section or an injection fan.

Sizing the return air damper will depend on the return/relief fan arrangement. *Figure 3-41* shows a constant volume system with a return fan provided

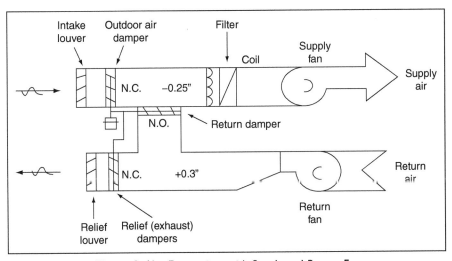

Figure 3-41 Economizer with Supply and Return Fan

for economizer relief. Typical plenum pressures are listed for illustration. The relief/exhaust plenum is pressurized by the return fan (+0.3 inches in this example) so that air can be exhausted through the relief damper and louver. The mixing plenum is under negative pressure (−25 inches in the example) so that air can be drawn into the plenum from the outdoors. Therefore, the return air damper is sized to absorb this pressure difference. In this example, it must be sized for a 0.55 inch pressure drop. Pressure drop will be a function of the drop through the damper itself and that due to the inlet and exit conditions. Refer to the *ASHRAE Fundamentals Handbook* or the *SMACNA HVAC Systems – Duct Design* (SMACNA, 1990) for inlet and outlet loss coefficients of various configurations. Control of these systems very much depends on proper sizing of all of the individual components that make up the air handling system, which can be further explored in upcoming chapters and in other references (Chen and Demster, 1996).

With variable air volume systems, the pressure in the relief/exhaust plenum often varies from a peak when the system is in 100% outdoor air (100% exhaust air) position to a minimum pressure, perhaps even neutral or slightly negative, when the system is only providing minimum outdoor air. The pressure variation will depend on how the return fan is controlled. A typical application where the return fan is sized to be equal to the supply fan less the minimum outdoor air intake rate is shown in *Figure 3-42* at both the 100% outdoor air and minimum outdoor air positions. The return air damper should be sized for the minimum outdoor air condition because that is when the return air damper is wide open. Note that the maximum air flow through the return damper is less than the total supply air flow rate by the minimum outdoor air quantity.

Figures 3-41 and *3-42* also show a relief (exhaust) damper. This is typically a motorized control damper, sized in the same manner as the outdoor air damper and interlocked with the outdoor air and return air dampers, so that the three dampers operate in unison. The interlock might be a physical linkage connection (as shown in the figures) but, on large systems, the damper will have its own actuator that operates over the same control range as the actuators controlling the outdoor air and return air dampers. The normal position of the relief damper is the same as the normal position of the outdoor air damper, usually normally closed. Always make sure that the minimum outside air ventilation rates are maintained.

Sometimes, a barometric backdraft damper is used instead of a motorized damper to reduce costs. This should only be done on systems that have volumetric fan tracking (VFT) capability (VFT is a control system that monitors the flow of air in the respective ducts, and makes adjustments to the dampers and other control devices in order to maintain balance) because the pressure in the relief plenum must vary as a function of how much air must be relieved, as it does in the example shown in *Figure 3-42*. Some research has shown that volumetric fan tracking is not recommended for minimum ventilation control. This variation will not occur naturally with constant volume systems, so barometric dampers should not be used in these applications. Barometric dampers also should not be used for systems located on top of a high-rise building because stack effect in cold weather will be sufficient to push the dampers open when the system is off. This causes air to exfiltrate out the damper with subsequent air infiltration to lower floors (Chen, 1996).

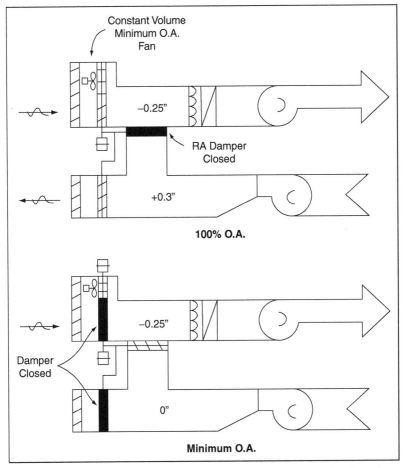

Figure 3-42 VAV System at 100% Outdoor Air and Minimum Outdoor Air

Figure 3-43 shows an economizer with a relief fan instead of a return fan. The pressures shown are typical of a ceiling plenum return air system with a very low pressure drop. In this application, the return air damper is typically sized to equalize the return air and outdoor air paths, so that the supply fan will see the same inlet pressure regardless of whether the system is operating in the 100% outdoor air or minimum outdoor air position. In this example, the damper would be sized for 0.1 inch pressure drop. Because the return path will usually have a higher pressure drop than the outdoor air path, the return damper is often made as large as practical (around 1000 fpm) and the outdoor air damper may then be reduced to equalize the pressure drops. Equalizing pressure drops is not necessary for VAV systems because any difference in pressure will be compensated for by the supply fan controls.

It is common for outdoor air and relief/exhaust air dampers to be of the low leakage type because a tight seal is desired when the system is off to reduce infiltration and exfiltration. This is particularly critical in cold climates where infiltration can cause damage such as coil freeze-ups. If the system runs

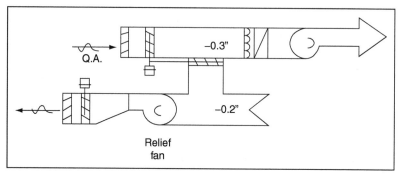

Figure 3-43   Economizer with Relief Fan

continuously (as it might in an institutional or manufacturing application), outdoor air leakage is not usually critical because a minimum outdoor air intake is typically required at all times for ventilation.

Often overlooked is the need to minimize leakage through the return air damper. When the system operates on 100% outdoor air, as it will when the outdoor air temperature is between the desired cooling supply air temperature set point and the economizer high limit condition, any leakage of return air into the mixing plenum will increase cooling energy usage. Therefore, a low leakage return air damper should be used for all economizer applications.

For more detailed information and worked examples on these damper arrangements see the ASHRAE Guideline 16-2003 Selecting Outdoor, Return, and Relief Dampers for Air-Side Economizer Systems.

The Next Step

In the next chapter, we will learn about sensors used to measure the controlled variable. Sensors commonly used in HVAC applications measure temperature, moisture content in air, differential pressure, flow, and current.

Bibliography

Alley, R. (1988) "Selecting and sizing outside and return air dampers for VAV economizer systems." *ASHRAE Transactions*. ASHRAE. Vol. 94, Pt. 1, pp. 1457–1466.

AMCA International 500-D, *Laboratory Methods of Testing Dampers for Rating*. Arlington Heights, IL: Air Movement and Control Association International, Inc.

ANSI/ASHRAE/IESNA Standard 90.1-2004 (2004) *Energy Standard for Buildings Except Low-rise Residential Buildings*. Atlanta, GA: American Society of Heating, Refrigerating and Air-conditioning Engineers, Inc.

ASHRAE Guideline 16-2003 (2003) *Selecting Outdoor, Return, and Relief Dampers for Air-side Economizer Systems*. Atlanta, GA: American Society of Heating, Refrigerating and Air-conditioning Engineers, Inc.

ASHRAE Handbooks:
Fundamentals 2005 – Fundamentals of Controls
HVAC Systems and Equipment 2004 – Valves
HVAC Applications 2007 – Testing, Adjusting, and Balancing.

ASHRAE (2004) Research Report (RP-1157) Flow Resistance and Modulating Characteristics of Control Dampers.

Avery, G. (1986) "VAV: an outside air economizer cycle." *ASHRAE Journal*. Vol. 28, No. 12, pp. 26–30.

Chen, S. and Demster, S. (1996) *Variable Air Volume Systems for Environmental Quality*. New York: McGraw-Hill. Chapter 5.

Hegberg, R. (1997) "Selecting control and balancing valves in a variable flow system." *ASHRAE Journal*. Vol. 39, No. 6, pp. 53–62.

Hegberg, R. (2001–2002) *Various Presentations from the Little Red Schoolhouse*. Chicago, IL: Bell and Gossett.

SMACNA. 1990. *HVAC Systems – Duct Design*. Chantilly, VA: Sheet Metal and Air Conditioning Contractors' National Association.

Chapter 4

Sensors and Auxiliary Devices

Contents of Chapter 4

Study Objectives of Chapter 4

Sensors are used to measure the controlled variable. Without measurement, there can be no control. Sensors are also used for monitoring purposes to keep the operator informed about elements in the system that indicate proper (or improper) operation. Sensors commonly used in HVAC applications include temperature, Carbon Dioxide (CO_2), Carbon Monoxide (CO), relative humidity, dewpoint, differential pressure, sensors used to estimate indoor air quality (IAQ), and velocity/flow sensors.

After studying this chapter, you should:

Be familiar with common temperature sensing devices.
Understand the psychrometric chart and moisture measurement in air.
Understand how moisture is measured and the effects of temperature on the measurement.
Understand how differential pressure is sensed.
Be aware of some of the sensors used for the estimation of IAQ.
Understand about different types of air and water flow sensors and the HVAC applications for which they are best suited.
Become familiar with auxiliary devices common to control systems.

Note that sensing technology is a rapidly changing field. Many of the physical sensors used in the past are being replaced by small electric/electronic devices which provide a vast array of qualities, reliabilities, and costs. The most common devices used in HVAC applications at the time of this publication are described below, but other devices are being used and new sensor technologies are being developed all the time. Manufacturers' catalogs and representatives should be consulted for the latest options. Remember that the latest may, or may not, be the best initial value or the best long-term value.

4.1 Introduction to Terms

We are going to start this chapter with some general discussion about terms, their use, and their definitions. Each definition is taken [B[from the *ASHRAE Terminology of Heating, Ventilating, Air Conditioning, and Refrigeration* (1991). You are not expected to learn them but rather to understand what they mean and how they are very important in some situations and not in others.

Accuracy

1. Conformity of an indicated value to an accepted standard value, or true value. Quantitatively, it should be expressed as an error or an uncertainty. The property is the joint effect of method, observer, apparatus, and environment. Accuracy is impaired by mistakes, by systematic bias such as abnormal ambient temperature, or by random errors (imprecision).
2. Degree of freedom from error, that is, the degree of conformity to truth or to a rule. Accuracy is contrasted with precision, e.g., four-place numbers are less precise than six-place numbers; nevertheless, a properly computed four-place number might be more accurate than an improperly computed six-place number.
3. Ability of an instrument to indicate the true value of a measured physical quantity.

The fundamental issue with accuracy is "How true is the value?"

Note the issue of apparent numerical accuracy and real accuracy. For example, compare a thermometer that provides readings in 1°F. If this thermometer is just 1°F in error it will provide more accurate readings than a thermometer that reads to 0.1°F which is 2°F out of true. In general, the number of digits is not a certain indication of accuracy. Accuracy is particularly important where sensors are replaced without field adjustment.

Range

1. Difference between the highest and the lowest operational values, such as pressure, temperature, rate of flow, or computer values.
2. Region between limits within which a quantity is measured, transmitted, or received, expressed by stating the lower and upper range values.

A sensor has a range of operation that is stated by the manufacturer. In general, one should avoid choosing sensors that will have to operate close to either end of their range.

Reliability

1. Mathematical probability that a device will perform its objective adequately for the period of time intended under the operating conditions specified.
2. Probability that a device will function without failure over a specified time period or amount of usage.
3. Probability that an instrument's repeatability and accuracy will continue to fall within specified limits.

Note that reliability covers two issues. First, "will it keep on working?" – what is the expected life. Second, "will it keep on working the same to the same standards over time?" To deal with significant drift over time one needs a maintenance system capable of regular checks on performance and of recalibrating the device.

Repeatability, Precision

1. Closeness of agreement among repeated measurements of the same variable under the same conditions.
2. Closeness of agreement among consecutive measurements of the output for the same value of input approaching from the same direction.

Repeatability covers the issue of getting the same reading under the same circumstances. You may have noticed the words "approaching from the same direction." This is more significant with some sensors than others. An example of this from an earlier chapter is the proportional controller which will have an offset in one direction in cooling mode and the opposite direction in heating mode.

Transmitter

A device that accepts a signal and transmits the information in the signal in a different form to another device. In many of the devices discussed in this chapter, the transmitter measures a change in resistance and, based on built in linear, or nonlinear, rules outputs signal proportional to the change in measured variable. This signal normally conforms to a 0–5 V, 0–10 V, or 4–20 milliamp (mA) signal. The voltage outputs work well for HVAC. The 4–20 mA signal is more robust, particularly for long cable lengths, and is largely used in industrial situations.

The accuracy of measurement is the combined accuracy of sensor and transmitter.

Figure 4-1 (Hegberg, 2001–2002) depicts a sensor which is precise and inaccurate, and a sensor that is imprecise and relatively accurate. The sensor that is precise and inaccurate provides an output with a permanent offset from the true value. Its smooth performance can give one a sense of "accuracy" that is unjustified. On the other hand, the imprecise and accurate unit jogs around in a most inconsistent manner, although it is providing a much closer-to-true value. The preference is always for the most precise and accurate sensor choices that the budget allows.

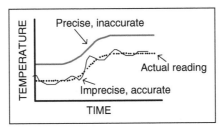

Figure 4-1 Accuracy and Precision (Hegberg, 2001–2002)

As of late, interchangeability and interoperability of sensors with differing manufacturers' control systems is becoming a more and more important feature to be considered in the buying and specifying process. Interchangeability deals with the ability to physically replace a sensor from one manufacturer with a sensor from another manufacturer, and have them read the same values. Interoperability deals with communication between the sensor and the system to which it is connected. A very simple example: there are standards for passing an electrical signal from sensor to system, 0–10 V and 4–20 mA. One cannot simply replace a 0–10 V unit with a 4–20 mA unit. When we get into direct digital controls (DDC) later in the course, the issue of interoperability becomes very important as there are many communications protocols by which devices communicate with systems.

In order to specify and design sensor systems, evaluation of "first cost" effectiveness such as qualitative items like installation time, accuracy, precision, reliability, repeatability, durability, maintenance, repair/replacement costs, compatibility, etc., should be considered. In too many cases in everyday practice, first costs often overwhelm the buying decision.

4.2 Temperature Sensors

In air-conditioning applications, temperature is typically the primary controlled variable. In comfort HVAC applications, temperature is used as the surrogate for human comfort because it is typically the primary factor affecting comfort. Other factors (such as humidity, air velocity, and radiant temperature) also affect comfort, but usually in a less significant way. Accordingly, these secondary factors are seldom directly measured and controlled, although they are generally indirectly controlled via the design of the HVAC system (in the case of humidity and air velocity), and building envelope (in the case of radiant temperature).

Temperature sensors can be categorized by the effect used to generate the temperature-versus-signal response:

Dimetal

Probably the first temperature sensor used for automatic control purposes was the bimetallic sensor (or bimetal for short), as shown in *Figure 4-2*.

This consists of two metal strips joined together continuously by welding or other means. The metals are selected so that each has a very different coefficient

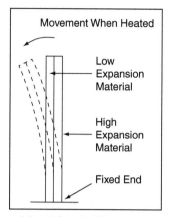

Figure 4-2   Bimetallic Temperature Sensor

of expansion (different rates of expansion relative to a change in temperature). Because one strip expands and contracts at a greater rate than the other, a change in temperature will cause the bimetal strip to bend, as shown in the figure. This action can be used in various control systems, both modulating and two-position. For instance, the strip can be mounted so that as the temperature rises (cooling) or falls (heating), the bending action will cause the strip to close a contact, completing an electrical circuit. To provide a firm closure, a small magnet is mounted to provide snap-action on opening and closing. In this example, the bimetal is serving as both a sensor and a controller. Bimetal strips are also used in sensing-only applications such as pneumatic controls. A typical temperature sensor has a bimetal strip which opens, or closes, a small orifice allowing more, or less, air through to the output device.

Another form of the bimetallic sensor, one that is only two-position, is shown in *Figure 4-3*.

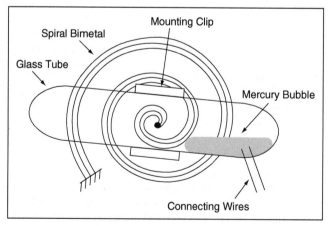

Figure 4-3   Mercury Switch

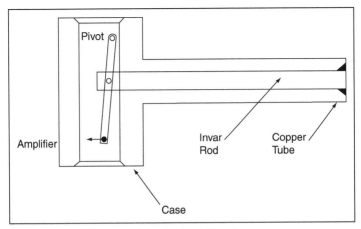

Figure 4-4   Rod-and-Tube Sensors

This spiral bimetal was widely used for simple two-position electric controls. The bimetal is fastened at the exterior end and a small glass mercury switch is mounted at the center of the spiral. As temperature change causes the spiral to wind or unwind, the mercury switch tilts and finally tips to the other position, causing the drop of mercury to make or break the circuit. When provided with an indicating pointer and a scale, the spiral bimetal can be used for temperature indication. This very common device was used in residential room thermostats. Due to the toxicity of free mercury these thermostats are no longer legally sold in many jurisdictions. With mercury being so toxic that old thermostats should not be discarded into the municipal waste system.

Figure 4-4 shows a rod-and-tube sensor, another type of bimetal sensor used for insertion into ducts or pipes. It consists of a tube made of a metal with a high coefficient of expansion enclosing a rod made of a low expansion material attached to the tube at one end. Changes in temperature then change the length of the rod and tube at different rates, causing the free end of the rod to move and expand and contract the amplifier plate, thereby generating the temperature signal.

Fluid Expansion

The bulb-and-capillary sensor (see *Figure 4-5*) utilizes a temperature-sensitive fluid contained in a bulb with a capillary connection to a chamber with a flexible diaphragm.

A change in temperature will cause a volume change in the fluid, which will cause the diaphragm to deflect. With the proper linkages, this can be used for either two-position or modulating control, in electric, electronic, or pneumatic systems. It is sometimes called a remote bulb sensor and is usually provided with fittings suitable for insertion into a duct, pipe, or tank. Averaging bulbs, where the temperature sensitive element is extended to sense more than a single point, are used where stratification is expected (such as after outdoor air/return air mixing boxes).

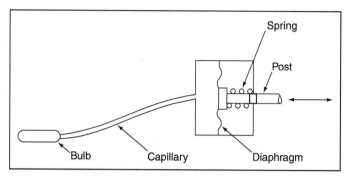

Figure 4-5   Capillary Sensors

Capillaries can be made to be temperature compensated to minimize the effect of the atmosphere through which the capillary passes. This is done by using two dissimilar metals for the capillary, one on the outside (usually stainless steel) and another metal on the inside with the sensing fluid in between. The materials are selected so that the differential coefficient of expansion of the two metals exactly equals the coefficient of expansion of the sensing fluid. Thus, as the capillary expands and contracts due to changes in ambient temperatures, it makes room for the fluid as it expands and contracts at the same rate. In this way, the ambient temperature changes do not affect the fluid pressure signal.

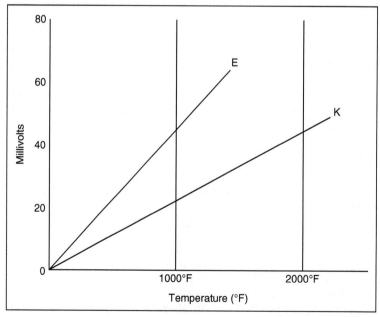

Figure 4-6   Thermocouples, E and K characteristics

The sealed bellows sensor (no figure shown) uses a similar principle. The bellows is filled with a gas or liquid with a high thermal expansion coefficient. Temperature changes cause the bellows to expand and contract, which can be measured to indicate temperature changes or can be used directly to move a controlled device.

Electrical, Self-powered

Thermocouples are formed by a junction of two dissimilar metals that develop a varying electromagnetic force (voltage) when exposed to different temperatures. For example, an iron wire and a constantan wire can be joined at their ends to form a junction. If the junction is heated to 100°F above ambient, about 3 mV will be generated at the hot junction. This same effect is used as a source of power in a thermopile which is many thermocouples connected together to produce a more powerful output. The most commonly used thermocouple materials (and their industry standard letter designations) are platinum-rhodium (Type S or R), chromel-alumel (Type K) (see *Figure 4-6*), copper-constantan (Type T), and iron-constantan (Type J). Accuracy for handheld instruments ranges from ±0.5°F to ±5°F for a calibrated thermocouple.

Thermocouples are inexpensive and commonly used for hand-held temperature sensors because they are small and able to reach steady-state quickly. *Table 4-1* depicts thermocouple advantages and disadvantages. Other than this application, thermocouples are used in HVAC applications for higher temperature measurement in boilers and flues. For general temperature other devices are used as they are more accurate and simpler to apply.

Electrical Resistance

Modern analog electronic and digital control systems generally rely on devices that resistance changes with temperature. Listed roughly in the order of commonality and popularity, these include thermistors, resistance temperature detectors (RTDs), and integrated circuit temperature sensors.

Thermistors are semiconductor compounds (*Figure 4-7*), that exhibit a large change in resistance, with changes in temperature usually decreasing as the temperature increases. The Y-axis of *Figure 4-7* is the ratio of resistance

Table 4-1   Thermocouple – Advantages and Disadvantages

Thermocouples	
Advantages	**Disadvantages**
Self-powered	Nonlinear
Simple	Reference required for accuracy
Rugged	Least stable
Fast response	Least sensitive
Wide variety	
Wide temperature range	
Inexpensive for lower accuracy	

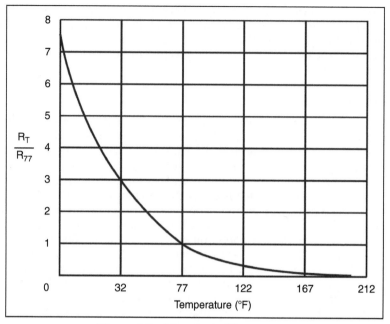

Figure 4-7   Thermistor Characteristic

compared to the resistance at 77°F. The characteristic resistance-temperature curve is nonlinear. The current, passed through the sensor to establish resistance, heats the sensor offsetting the reading to some extent (called self-heating). In electronic applications, conditioning circuits are provided in the transmitter to create a linear signal from the resistance change. In digital control systems, the variable resistance is often translated to a temperature signal by using a software look-up table that maps the temperature corresponding to the measured resistance, or by solving an exponential equation using exponents and coefficients provided by the thermistor manufacturer. Their main advantages and disadvantages are tabulated in *Table 4-2*.

Table 4-2   Thermistor – Advantages and Disadvantages

Thermistor	
Advantages	**Disadvantages**
High resistance change	Nonlinear
Fast response	Fragile
Two-wire measurement	Current source required
Low cost	Self-heating

Thermistors typically have an accuracy around ±0.5°F, but they can be as accurate as ±0.2°F. They have a high sensitivity, in other words they have a fast and detailed response to a change in temperature. However, they drift over time, and regular calibration is required to maintain this accuracy. At one time, calibration was required about every six months or so, but the quality of thermistors has improved in recent years, reducing the frequency interval to once every five years or more. For instance, commercial grade thermistors are now available with a guaranteed maximum drift of 0.05°F over a five-year period. They now have long-term stability and a fast response at a low cost.

The RTD is another of the most commonly used temperature sensor in analog electronic and digital control systems because it is very stable and accurate, and advances in manufacturing techniques have rapidly brought prices down. As the name implies, the RTD is constructed of a metal that has a resistance variation as a direct acting function of temperature that is linear over the range of application (*Figure 4-8*). Common materials include platinum, copper-nickel, copper, tungsten, and some nickel-iron alloys. In HVAC applications, RTDs are often in a wound wire configuration, with the RTD metal formed into a fine wire and wrapped around a core. Coil wound RTDs cost more than thermistors but they are more stable, so regular recalibration is not usually required. Standard platinum RTDs have a reference resistance of 100 ohms at 0°C. This low resistance (compared to 10,000–100,000 ohms for thermistors) typically requires that the measurement circuit compensate for or eliminate the resistance of the wiring used to connect the RTD to the detector, because this resistance will be on the same order of magnitude as the RTD. To do this either the detector must be calibrated to compensate

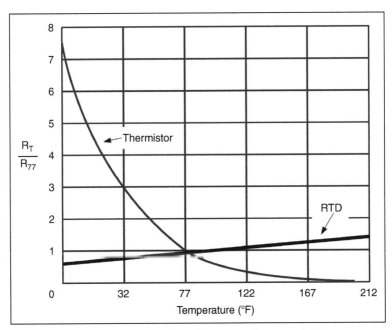

Figure 4-8   Thermistor and RTD Resistance Change with Temperature

for wiring resistance or, more commonly, three-wire or four-wire circuits are used that balance or eliminate wiring resistance. For HVAC applications, platinum RTDs rated at 100 ohms are typically about ±0.5°F at the calibration point to ±1.0°F accuracy over the application range. However, high purity platinum sensors can have an accuracy of ±0.02°F or even better.

A recent development is the thin-film platinum RTD, which has a reference resistance of about 1000 ohms. Made by deposition techniques that substantially reduce the cost, these sensors are one of the primary reasons why RTDs began to replace thermistors in electronic and digital control systems. Thin-film RTDs have accuracies on the order of ±0.5°F to ±1.0°F at their calibration point. As the units are dependent on the behavior of platinum metal they have a very, very low drift. The main advantages and disadvantages of RTDs are shown in *Table 4-3*.

Integrated circuit (IC) temperature sensors (also called solid-state temperature sensors or linear diodes) are based on semiconductor diodes and transistors that exhibit reproducible temperature dependence. They are typically sold as ready-made, packaged integrated circuits (sensor and transmitter) with built-in conditioning to produce a linear resistance to temperature signal. Solid-state sensors have the advantage of requiring no calibration, and their cost and accuracy are on the order of thin-film platinum RTDs. See *Table 4-4* for their main advantages and disadvantages.

The process to select an appropriate sensor type should be concerned with the economics, accuracy, and long-term reliability of the sensor. A summary of sensor characteristics is shown in *Table 4-5*. In HVAC systems, for the

Table 4-3   RTD – Advantages and Disadvantages

RTDs	
Advantages	**Disadvantages**
Most stable	Expensive
Most accurate	Current source required
Most linear	Coiled type low resistance, 100 ohms, requires good temperature compensation
	Film type has relatively low resistance
	Self-heating

Table 4-4   Linear Diodes – Advantages and Disadvantages

Linear Diodes	
Advantages	**Disadvantages**
Most linear	Use up to 330°F
Inexpensive	Power supply required
	Slow
	Self-heating
	Limited configurations

Table 4-5   Summary of Sensors

Temperature Sensors Comparison				
Type	Primary Use	Advantages	Disadvantages	Response Time
Thermocouple	Portable units and high temperature use < 5000°F	Inexpensive Self-powered for average accuracy	Very low voltage output	Slow to fast depending on wire gauge
Thermistor	High sensitivity General use < 300°F	Very large resistance change	Nonlinear Fragile Self-heating	Fast
RTD	General purpose < 1400°F	Very accurate Interchangeable Very stable	Relatively expensive	Long for coil Medium/ fast for foil Short for thin film
Integrated circuit	General purpose < 400°F	Linear output Relatively inexpensive	Not rugged Limited selection	Medium/ fast

most part, extremely accurate devices are not usually needed to produce the required actions. All of the above sensor types are within this acceptable window of requirements. Different manufacturers of controls typically carry the capability to use any of these sensors.

If extreme accuracy or extreme reliability is required, specify these requirements and highlight them in the design specifications.

The final commissioning process is critical, and necessary for the assurance of proper control system and sensor performance to the process.

The useful accuracy of temperature sensors varies considerably. Early in this text it was mentioned that room temperature sensors need to be reliable but not accurate if the occupant can adjust them. The occupant will adjust the thermostat to their comfort and accuracy of calibration in degrees Fahrenheit is not the issue.

Now consider an air-conditioning plant which includes an air economizer (uses outside air for cooling when appropriate) and a cooling coil. The plant supplies air at a constant temperature of 54°F. There are two temperature sensors we are going to consider: outside air and supply air. The outside air sensor is used for information and controls when the change is made from 100% outside air to minimum outside air.

From the point of view of plant performance, the outside air temperature matters at the changeover point but not elsewhere. Any nonlinearity would be irrelevant as long as it is set correctly at the changeover point. Even at the changeover point, an error only matters for a relatively few hours in the year in most climates.

Now consider the performance of the supply air temperature sensor. It is to maintain 54°F and let us suppose the return air temperature is 75°F. The cooling effect is the air being heated in the spaces from 54°F to 75°F, a temperature rise of 21°F. Now let us suppose the temperature sensor is just 1°F off and the supply temperature is 55°F. The cooling capacity is down from 55°F to 75°F, or 20°F, a drop of 5% in cooling capacity. An error of 2°F produces a reduction of 10%. Accuracy really does matter here. However, the accuracy is only needed at 54°F not at higher or lower temperatures.

This issue of accuracy is particularly important where sensors are replaced without site calibration. A higher supplied accuracy is needed in this change-it-out situation without site calibration.

4.3 Humidity and the Psychrometric Chart

Humidity is the moisture content in air. Before we consider humidity sensors, it is important that you understand what is being measured and how the various measurements relating to moisture content and temperature interact. Since moisture and temperature relate to the energy, or enthalpy, of the moist air we will also introduce that issue.

The relationships between temperature, moisture content, and energy are most easily understood using a visual aid called the *"psychrometric chart."*

The psychrometric chart is an industry-standard tool that is used to visualize the interrelationships between dry air, moisture, and energy. If you are responsible for the design or maintenance of any aspect of air-conditioning in buildings, a clear and comfortable understanding of the chart will make your job easier.

Initially, the chart can be intimidating, but as you work with it you will discover that the relationships that it illustrates are relatively easy to understand. Once you are comfortable with it, you will discover that it is a tool that can make it easier to troubleshoot air-conditioning problems in buildings. In this course, we will only introduce the psychrometric chart and provide a very brief overview of its structure.

The psychrometric chart is built upon two simple concepts.

1. Indoor air is a mixture of dry air and water vapor.
2. There is a specific amount of energy in the mixture at a specific temperature and pressure.

Indoor Air is a Mixture of Dry Air and Water Vapor

The air we live in is a mixture of both dry air and *water vapor*. Both are invisible gases. The water vapor in air is also called *moisture* or *humidity*. The quantity of water vapor in air is expressed as *"pounds of water vapor per pound of air."* This ratio is called the "humidity ratio," abbreviation W, and the units are pounds of water/pound of dry air, lb_w/lb_{da}, often abbreviated to lb/lb.

The exact properties of moist air vary with pressure. As pressure reduces as altitude increases the properties of moist air change with altitude. Typically, psychrometric charts are printed based on standard pressure at sea level. For the rest of this section we will consider pressure as constant.

To understand the relationship between water vapor, air, and temperature, we will consider two conditions:

a. The air temperature is constant, but the quantity of water vapor is increasing.

b. The air temperature is dropping, but the quantity of water vapor is constant.

a. The temperature is constant, but the quantity of water vapor is increasing. If the temperature remains constant, then as the quantity of water vapor in the air increases, the humidity increases. However, at every temperature point, there is a maximum amount of water vapor that can co-exist with the air. The point at which this maximum is reached is called *the saturation point*. If more water vapor is added after the saturation point is reached, then an equal amount of water vapor condenses and takes the form of either water droplets or ice crystals.

Outdoors, we see water droplets in the air as fog, clouds, or rain and we see ice crystals in the air as snow or hail. The psychrometric chart only considers the conditions up to the saturation point; therefore, it only considers the effects of water in the vapor phase and does not deal with water droplets or ice crystals.

b. The temperature is dropping, but the quantity of water vapor is constant. If the air is cooled sufficiently, it reaches the *saturation line*. If it is cooled even more, moisture will condense out and dew forms.

For example, if a cold canned drink is taken out of the refrigerator and left for a few minutes, the container gets damp. This is because the moist air is in contact with the chilled container. The container cools the air that it comes in contact with to a temperature that is below saturation, and dew forms. This temperature, at which the air starts to produce condensation, is called the *dew-point temperature*.

Relative Humidity

Figure 4-9 is a plot of the maximum quantity of water vapor per pound of air against air temperature. The X-axis is temperature. The Y-axis is the proportion of water vapor to dry air, measured in pounds of water vapor per pound of dry

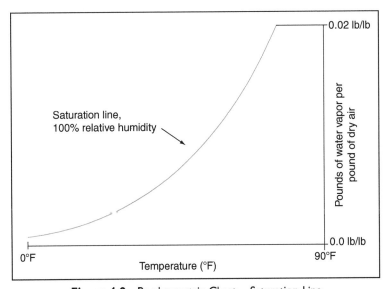

Figure 4-9 Psychrometric Chart – Saturation Line

air. The curved "maximum water vapor line" is called the "saturation line." It is also known as *100% relative humidity*, abbreviated to *100% rh*. At any point on the saturation line, the air has 100% of the water vapor per pound of air that can coexist with dry air at that temperature.

When the same volume of air contains only half the weight of water vapor that it has the capacity to hold at that temperature, we call it *50% relative humidity* or *50% rh*. This is shown in *Figure 4-10*. Air at any point on the 50% rh line has half the water vapor that the same volume of air could have at that temperature.

As you can see on the chart, the maximum amount of water vapor that moist air can contain increases rapidly with increasing temperature. For example, moist air at the freezing point, 32°F, can contain only 0.4% of its weight as water vapor. However, indoors, at a temperature of 72°F, the moist air can contain nearly 1.7% of its weight as water vapor – over four times as much.

Consider *Figure 4-11*, and this example:

On a miserable wet day it might be 36°F outside, with the air rather humid, at 70% relative humidity. Bring that air into your building. Heat it to 70°F. This brings the relative humidity down to about 20%. This change in relative humidity is shown in *Figure 4-12*, from *Point 1 → 2*. A cool damp day outside provides air for a dry day indoors! Note that the absolute amount of water vapor in the air has remained the same, at 0.003 pounds of water vapor per pound of dry air, but as the temperature rises, the relative humidity falls.

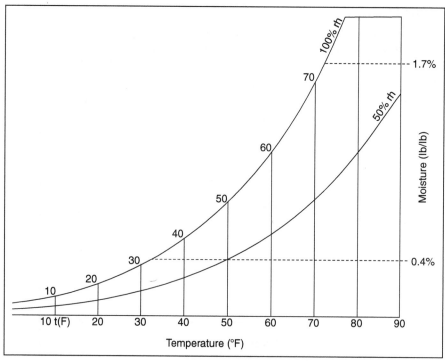

Figure 4-10 Psychrometric Chart – 50% Relative Humidity Line

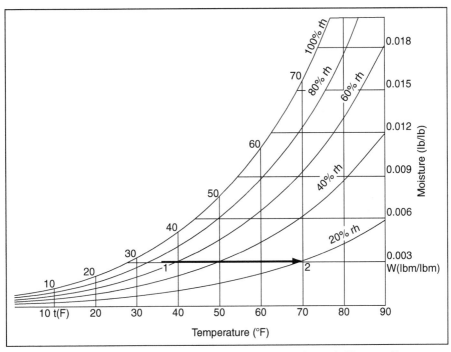

Figure 4-11   Psychrometric Chart – Change in Relative Humidity with Change in Temperature

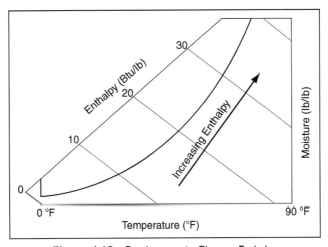

Figure 4-12   Psychrometric Chart – Enthalpy

Here is an example for you to try, using *Figure 4-11*.

Suppose it is a warm day with an outside temperature of 90°F and relative humidity at 40%. We have an air-conditioned space that is at 73°F. Some of the outside air leaks into our air-conditioned space. This leakage is called *infiltration*.

Plot the process on *Figure 4-12*.

1. Find the start condition, 90°F and 40% rh, moisture content 0.012 lb/lb.
2. Then cool this air: move left, at constant moisture content to 73°F.
3. Notice that the cooled air now has a relative humidity of about 70%.

Relative humidity of 70% is high enough to cause mold problems in buildings. Therefore, in hot moist climates, to prevent infiltration and mold generation, it is valuable to maintain a small positive pressure in buildings.

There is a specific amount of energy in the air mixture at a specific temperature and pressure. This brings us to the second concept that the psychrometric chart illustrates. There is a specific amount of energy in the air water-vapor mixture at a specific temperature. The energy of this mixture is dependent on two measures:

- The temperature of the air.
- The proportion of water vapor in the air.

There is more energy in air at higher temperatures. The addition of heat to raise the temperature is called adding *"sensible heat."* There is also more energy when there is more water vapor in the air. The energy that the water vapor contains is referred to as its *"latent heat."*

The measure of the total energy of both the sensible heat in the air and the latent heat in the water vapor is commonly called "enthalpy." Enthalpy can be raised by adding energy to the mixture of dry air and water vapor. This can be accomplished by adding either or both:

- sensible heat to the air
- more water vapor, which increases the latent heat of the mixture.

On the psychrometric chart, lines of constant enthalpy slope down from left to right as shown in *Figure 4-13* and are labeled "Enthalpy."

The zero is arbitrarily chosen as zero at 0°F and zero moisture content. The unit measure for enthalpy is *British Thermal Units per pound of dry air*, abbreviated as *Btu/lb*.

Heating: The process of heating involves the addition of sensible heat energy. *Figure 4-13* illustrates outside air at 47°F and almost 90% relative humidity that has been heated to 72°F. This process increases the enthalpy in the air from approximately 18–24 Btu/lb. Note that the process line is horizontal because no water vapor is being added to, or removed from the air – we are just heating the mixture. In the process, the relative humidity drops from almost 90% rh down to about 36% rh.

Here is an example for you to try.

Plot this process on *Figure 4-14*.

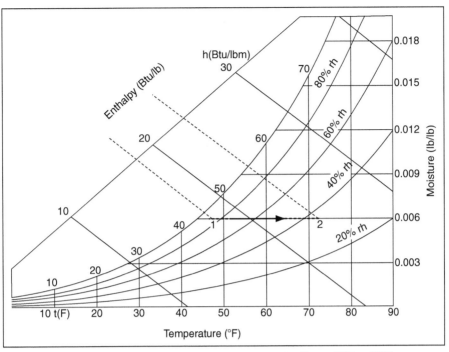

Figure 4-13   Psychrometric Chart – Heating Air From 47°F to 72°F

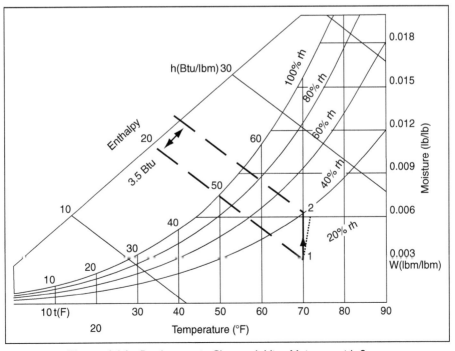

Figure 4-14   Psychrometric Chart – Adding Moisture with Steam

Suppose it is a cool day with an outside temperature of 40°F and 60% rh. We have an air-conditioned space and the air is heated to 70°F. There is no change in the amount of water vapor in the air. The enthalpy rises from about 13 Btu/lb to 20 Btu/lb, an increase of 7 Btu/lb.

As you can see, the humidity would have dropped to 20% rh. This is quite dry so let us assume that we are to raise the humidity to a more comfortable 40%. As you can see on the chart, this raises the enthalpy by an additional 3.5 Btu/lb.

Humidification: The addition of water vapor to air is a process called *"humidification."* Humidification occurs when water absorbs energy, evaporates into water vapor, and mixes with air. The energy that the water absorbs is called *"latent heat."*

There are two ways for humidification to occur. In both methods, energy is added to the water to create water vapor.

1. **Water can be heated.** When heat energy is added to the water, the water is transformed to its gaseous state, steam, which mixes into the air. In *Figure 4-14*, the vertical line, from Point 1 to Point 2, shows this process. The heat, energy, 3.5 Btu/lb, is put into the water to generate steam (vaporize it), which is then mixed with the air.

   In practical steam humidifiers, the added steam is hotter than the air and the piping loses some heat into the air. Therefore, the air is both humidified and heated due to the addition of the water vapor. This combined humidification and heating would be shown by a line which slopes up and a little to the right in *Figure 4-14*.

2. **Let the water evaporate** into the air by spraying a fine mist of water droplets into the air. The fine water droplets absorb heat from the air as they evaporate. Alternatively, but using the same evaporation process, air can be passed over a wet fabric, or wet surface, enabling the water to evaporate into the air.

   In an evaporative humidifier, the evaporating water absorbs heat from the air to provide its latent heat for evaporation. As a result, the air temperature drops, as it is humidified. The process occurs with no external addition or removal of heat. It is called an *adiabatic process*. Since, there is no change in the heat energy (enthalpy) in the air stream, the addition of moisture, by evaporation, occurs along a line of constant enthalpy.

   Figure 4-15 shows the process. From Point 1, the moisture evaporates into the air and the temperature falls to 56°F (Point 2). During this evaporation, the relative humidity rises to about 65%. To reach our target of 70°F and 40% rh we must now heat the moistened air at Point 2 from 56°F to 70°F (Point 3) requiring 3.5 Btu/lb of dry air.

To summarize, we can humidify by adding heat to water to produce steam and mixing the steam with the air, or we can evaporate the moisture and heat the moistened air. We achieve the same result with the same input of heat by two different methods.

It has become much easier to control humidity in buildings but do be aware of the consequences. In a cold climate, maintaining higher humidity has a day-to-day energy cost. If humidity is maintained too high for the building, serious damage from condensation on the inside can occur. Within the walls, ice can cause serious structural damage to the exterior wall facing. In the humid

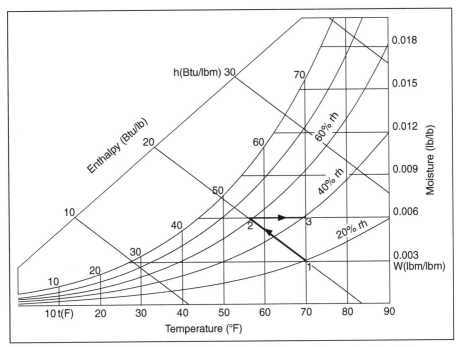

Figure 4-15   Psychrometric Chart – Adding Moisture, Evaporative Humidifier

climate, dehumidification is costly but failure to continuously limit the maximum humidity can lead to mold problems resulting in building closure. The Kalia Tower Hilton, Hawaii, mold problem involved closing the 453-room hotel for refurbishing at a cost of over $US50 million.

One last issue is the term wet-bulb temperature. We have discussed the fact that moisture evaporating into air cools the air. This property is used to obtain the wet bulb temperature. If a standard thermometer has its sensing bulb covered in a little sock of wet cotton gauze, and air blows quickly over it, the evaporation will cool the thermometer. An equilibrium temperature is reached which depends on the dry-bulb temperature and relative humidity. If the air is very dry, evaporation will be rapid and the cooling effect large. In saturated air the evaporation is zero and cooling zero, so dry-bulb temperature equals wet-bulb temperature at saturation.

Lines of constant wet-bulb temperature can be drawn on the psychrometric chart. They are almost parallel to the enthalpy lines and the error is not significant in normal HVAC except at high temperatures and low relative humidity.

If, for example, the dry-bulb temperature was 60°F and wet-bulb was 50°F, we can plot these on the chart as shown in *Figure 4-16* and find the relative humidity to be 50%. If the temperature were 70°F and wet bulb still 50°F the relative humidity would be down at about 20%. Remember, the greater the wet-bulb temperature depression the lower the relative humidity.

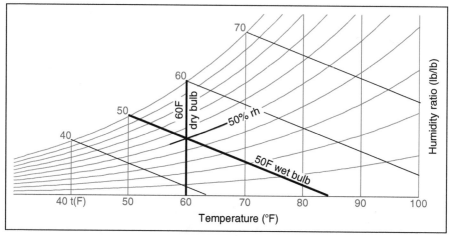

Figure 4-16 Plot of Dry-bulb and Wet-bulb Temperatures

This has been a very brief introduction to the concepts of the psychrometric chart. A typical published chart looks complicated as it has all the lines printed, but the simple underlying ideas are:

- Indoor air is a mixture of dry air and water vapor.
- There is a specific amount of total energy, called enthalpy, in the mixture at a specific temperature, moisture content, and pressure.
- There is a maximum limit to the amount of water vapor in the mixture at any particular temperature.

Now that we have an understanding of the relationships of dry air, moisture, and energy, at a particular pressure let us consider relative humidity, dew-point, and enthalpy sensors.

Figure 4-17 shows a section of a simple building with an air-conditioning unit drawing return air from the ceiling plenum and supplying to three spaces, A, B, and C. Each space has individual temperature control with its own thermostat and heater. The air handling unit has a relative humidity sensor in the middle space B. Assuming similar activities and the same temperature in each room the relative humidity will also be the same in each room.

Now let us assume that room A occupant likes it warmer. What will happen to the relative humidity in space A? Go down, up, or stay the same? Yes, it will go down. So the obvious thing to do is to average the relative humidity.

We can achieve this by moving the relative humidity sensor to the inlet of the air handling unit. If the occupants had a relative humidity sensor on their desk they could correctly complain that the relative humidity is going up and down. However, the control system records would show that system is maintaining the humidity perfectly constant. Both are right, how can this be?

The lights produce heat that heats the return air above the ceiling. During the night the lights are off so the return air from the rooms is at the same temperature as the return air into the unit. In the daytime, when the lights are on, the return air is heated in the plenum by the lights and the relative humidity

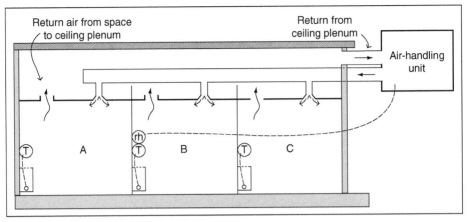

Figure 4-17   Building Arrangement

drops. The air handling unit compensates for this by raising the moisture content. This raises the humidity level in the spaces while keeping the relative humidity constant at the air-handler intake.

Now let us imagine that the roof of this building is not perfectly insulated. When the sun shines on the roof, the heat from the sun will also heat the plenum. This will also cause the relative humidity in the return air to go down and the air-handler will respond by raising moisture content in the system.

In this type of situation, it is a bad idea to use return air relative humidity to control multiple spaces. It is better to either use one space as the master space and maintain its humidity, or to use a dew-point sensor. The dew-point sensor can be mounted anywhere in the system as it is not influenced by the dry-bulb temperature. With the dew-point sensor in the return air it sensor will average the moisture content in all the spaces.

Since the relative humidity varies with temperature it is better to specify the control range for a space in terms of dry-bulb temperature and dew-point (or humidity ratio) rather than relative humidity.

Use of Enthalpy Sensors: Enthalpy sensors can be effective at reducing cooling costs, particularly in humid climates. Consider a system with an air economizer. The question to be answered is: "When should the system stop using 100% outside air and revert to minimum outside air?" Consider a plant where the return air is at 75°F and 50% relative humidity. This is shown on *Figure 4-18*. The enthalpy for this return air is also shown in bold. If the system uses temperature to make the decision, a temperature of 65°F would ensure that the switch was made before the outside air enthalpy rises above the return enthalpy, except for the very occasional possibility that the humidity is over 90%. This temperature setting avoids ever bringing in air with a higher enthalpy than the return air. The conditions for using return air are shown by the hatched area, all temperatures below 65°F and any moisture content.

Switching at 65°F ensured that excessively high enthalpy outside air is virtually never used but it also switches the plant well before it needs to in many situations. If, instead the switch is made based on enthalpy one has two

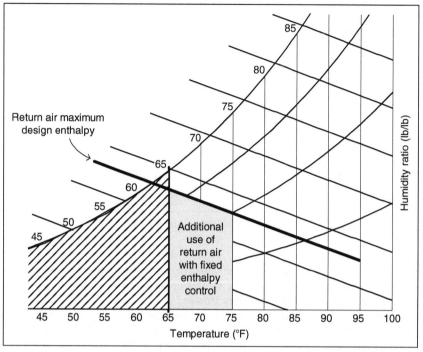

Figure 4-18 Temperature Versus Enthalpy for Switching off Economizer

choices. One could use a single enthalpy sensor set at the design return air enthalpy (the bold line in *Figure 4-18*). This would allow outside air to be used in the additional shaded area.

Better, would be to use an enthalpy sensor in the return air and outside air and make the decision to drop to minimum outside air when the outside air enthalpy rose to the return air enthalpy.

4.4 Moisture Sensors

Relative Humidity Sensors

Accurate, stable, and affordable humidity measurement has always been challenging to achieve in HVAC systems. ASHRAE Standard 62.1 Ventilation for Acceptable Indoor Air Quality requires maintaining inside relative humidity levels below 65%. Modern solid-state technology has improved this process, but, with any humidity sensor, periodic calibration and maintenance are required for sustained accuracy. For this reason, the designer must weigh the benefits of humidity measurement costs with the life expectancy and dependability, as well as the maintenance costs and the potential problems caused by imprecise measurements. It is usually desired to purchase the most dependable and cost effective sensors as feasible for the application.

Humidity may be sensed as relative humidity, dew-point temperature, or wet-bulb temperature, with relative humidity being by far the most common. The three parameters are all interrelated and the measurement of any one, along with coincident dry-bulb temperature (and barometric pressure if it varies significantly), can be used to determine any of the others using known psychrometric relationships and be properly used in control applications.

Relative Humidity (rh): Historically, the first humidity sensors used hygroscopic materials (materials that can absorb water vapor from the air) that change dimension in response to changes in humidity. These include animal hair, wood, and various fabrics, including some synthetic fabrics such as Nylon® and Dacron®. These mechanical sensors are still commonly used in portable sensors, such as you might have on your desk, inexpensive electric controls (humidistats), and inexpensive enthalpy economizer controllers. Their accuracy is generally no better than ±5% relative humidity, when new, due to variations in material quality and hysteresis effects.

Resistance-type humidity sensors use hygroscopic materials whose electrical resistance varies in a repeatable fashion when exposed to air of varying humidity. One type uses a sulfonated polystyrene resin placed on an insoluble surface. An electrically conductive layer is then bonded to the resin. The electrical resistance of the assembly varies nonlinearly but fairly repeatably with humidity. A linear signal is created using techniques similar to those used for thermistors described above. Like many humidity sensors, the accuracy of the resistance-type sensor can be severely affected if the surface is contaminated with substances (such as oil) that affect the water vapor absorption or desorption characteristics of the resin.

Capacitance-type humidity sensors are available in various forms, all based on the variation in electrical capacitance of a hygroscopic material. One type consists of an aluminum strip deposited with a layer of porous aluminum oxide underneath a very thin layer of gold. The aluminum and gold form the plates of the capacitor, with the aluminum oxide as the dielectric. The capacitance varies as a function of the water vapor absorbed in the aluminum oxide layer. Water vapor is absorbed and desorbed by passing through the very thin layer of gold. The accuracy of this sensor (called a Jason-type hygrometer) is very good up to 85% relative humidity but the sensor can become permanently damaged if exposed to higher humidity air and particularly condensation.

A variation of the Jason-type capacitance sensor uses a thin film of polymer in place of the aluminum oxide, *Figure 4-19.* The polymer is carefully selected to provide a capacitance change as a function of humidity but without the 85% humidity limitation of aluminum oxide. These sensors are available with accuracies ranging from ±5% rh to as fine as ±1% rh, including hysteresis and calibration uncertainty. Because of their accuracy and reliability, these sensors are becoming the style most commonly used in analog electronic and digital control systems. However, they can be expected to drift on the order of 1% to 3% rh per year under normal applications, so as with all humidity sensors, periodic and regular calibration (quarterly) is suggested.

Lithium Chloride Dew-point Sensors: Dew-point sensors are the most accurate type of humidity sensor, but they are also the most expensive. One type of dew-point sensor uses a saturated salt solution (usually lithium chloride) in contact with the air whose humidity is to be measured. When steady state is

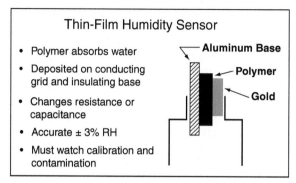

Figure 4-19   Thin-Film Sensor Example (Hegberg, 2001)

reached, the temperature of the solution is indicative of the dew-point of the air. This type of sensor is very accurate, very slow to respond, but inaccurate at low humidity levels. It is also sensitive to contamination and requires periodic maintenance and calibrations, and is relatively expensive. Newer versions use lithium chloride solution on a grid with an integral heater. The lithium chloride is hygroscopic and attracts moisture lowering the grid resistance. This lowering resistance increases the heater output which lowers the resistance. The balance between wetter lower resistance and heater higher resistance provides the signal for the dew-point. The accuracy of the sensor can be better than ±2.5°F.

Chilled-Mirror Dew-point Sensors: Another very accurate dew-point sensor is the chilled-mirror type (*Figure 4-20*).

In these devices, a sample of air flows through a small sensor chamber equipped with a light source, two photocells, and a chilled mirror. Light reflects off the mirror toward one photocell. When condensation forms on the mirror, the light is scattered rather than reflected directly. The system recognizes this by comparison with the direct reference photocell. The reduction in the light level indicates the presence of condensation on the mirror. At the moment of

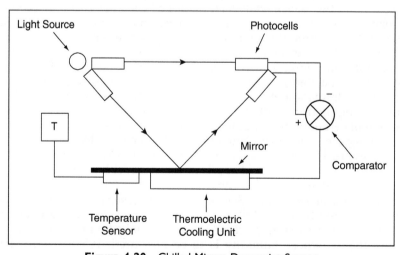

Figure 4-20   Chilled-Mirror Dew-point Sensor

condensation, an RTD temperature sensor records the surface temperature of the mirror. That surface temperature is the dew-point of the air flowing across the mirror.

The surface of the mirror is chilled by an array of semiconductors known as Peltier junctions, which form a thermoelectric cooler that can be controlled by varying its electrical current. A control circuit modulates the current passing through the semiconductors, keeping the temperature of the mirror constant at the dew-point of the air. Chilled-mirror sensors are not as widely used in commercial buildings as relative humidity sensors, mostly because of their cost and maintenance requirements. But they are highly accurate, and are often used to calibrate lower cost devices. Advantages include:

- They measure the dew-point directly. Controls can be set based on the instrument's output signal without the need to calculate the dew-point based on temperature and rh.
- They are accurate. Typical dew-point tolerance is $\pm 0.4°$F.
- There is a wide range. Even with a single stage of mirror cooling, the close tolerance measurement can easily be maintained between temperatures of 0 and 100°F with coincident dew-points between -20 and 80°F dry bulb.

Limitations of chilled-mirror sensors include:

- The cost. The lowest cost versions are about twice as expensive as close-tolerance rh sensors, and the broader-range chilled-mirror devices cost about five times more than the lowest cost versions.
- Contamination. The mirror surface must be kept clean and free of hygroscopic dust that creates condensation at a temperature *higher* than the true dew-point temperature of the air. The air sample must be filtered, and the filter must be replaced regularly when particulate loading is especially heavy.

Nevertheless, where very precise and repeatable humidity measurements are required, chilled-mirror sensors are a good choice.

Figure 4-21 shows different packaging of electrical signal sensors for different applications available in the marketplace. Note that the packaging for temperature sensors looks (and often is) almost the same.

Psychrometers: A psychrometer measures humidity by taking both a wet-bulb and a dry-bulb temperature reading. With those two values known, the other properties of the air, including its moisture content, can be determined

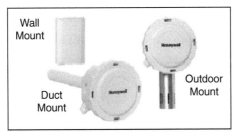

Figure 4-21   Examples of Humidity Sensors

by computation or by reading a psychrometric chart. In commercial buildings, psychrometers are seldom if ever used for control, but they are occasionally used to check the calibration of humidistats or relative humidity sensors.

Sling psychrometers are a choice for that purpose. These units consist of two thermometers with thin bulbs. One is covered in a cotton sleeve which is wetted with (ideally distilled) clean water. The two thermometers are mounted in a sling which is swung rapidly around-and-around and then quickly read to obtain a steady wet- and dry-bulb temperature. Be careful to use the sling psychrometer correctly as it does have some drawbacks. Slow air velocity, inadequate water coverage of the wick, radiation heating of the wet bulb, and contamination of the wet wick are compounded by the difficulty of being sure the wet-bulb reading is at its minimum while the thermometer is swinging. These problems mostly come from not slinging long enough to get down to wet-bulb steady-state, so the measurement error is always *above*, rather than below the true wet-bulb reading. In other words, poor measurements from sling psychrometers will always *overestimate* the true moisture content. The only exception occurs when cold water rather than ambient-temperature water wets the wick. In that case, it is possible to underestimate the true humidity level by taking a reading before stable conditions have been achieved.

Aspirated (fan powered) psychrometers with clean wet wicks using distilled water are more accurate than sling-type units. An aspirated psychrometer combines low cost with the fundamental measurement principle of wet- and dry-bulb readings. For typical humidity ranges of commercial buildings (30–60% rh at 68–75°F) aspirated psychrometers provide a reliable, low-cost way to check readings from low-accuracy sensors.

In an aspirated psychrometer, the wet- and dry-bulb thermometers are mounted inside a plastic case, which contains a battery-powered fan. The fan draws air across both dry and wet thermometers at a constant, high velocity to provide uniform evaporation. The case prevents radiation from influencing the temperature of the thermometer bulbs. The wick must be changed regularly *with gloved hands* to prevent skin oils and air stream particulate from affecting evaporation, and only ambient-temperature distilled water can be used to wet the wick. Further, the wick must remain completely wetted until the wet-bulb temperature has stopped dropping. As long as all those precautions are followed, aspirated psychrometers can be useful to cross-check readings from low-accuracy sensors. The advantages of aspirated psychrometers include the following:

- Recalibration is not an issue, as it is with electronic units, since physical properties are being directly measured.
- Reasonable accuracy in indoor environments. A tolerance of +5% of the wet-bulb reading can be achieved in careful operation in middle- and upper-range humidity levels.
- Portable. The instrument can be brought to a room sensor location easily.

The limitations of wet-bulb readings must also remain clear:

- Requires a psychrometric chart. To obtain humidity values, the operator must carefully plot the point and read values on an accurate psychrometric chart. Plotting and reading introduce two major sources of error. Poor results from aspirated psychrometers usually come from incautious plotting and reading of the psychrometric chart after the wet-bulb and dry-bulb readings are obtained. But most psychrometers do have charts already engraved on their bodies.

- Difficult to use in ducts. The device must draw air only from the duct and not from the air outside that duct. It is difficult to avoid air mixing when opening an access door, and difficult to read the results inside a dark duct.
- Difficult to use in low-relative-humidity air. Wet-bulb temperature readings below the freezing point of water are difficult to obtain because it takes a long time to cool the wick low enough to freeze the water, and a long time to stabilize the temperature after an ice layer has formed. These precautions are seldom taken outside of a carefully controlled laboratory test rig. That means psychrometers are seldom useful in low-humidity air streams where sub-freezing wet-bulb temperatures are common.
- Subject to error in reading the thermometers. For accurate results, the operator cannot neglect to define what fraction of a degree the thermometer is sensing. Reading fractions of a degree from small thermometers requires care, good light, and good eyesight.
- Subject to errors of contamination. In the day-to-day reality of building operations, the wet-bulb wick is not always kept clean of particulate, and is often wetted with mineral-laden water or handled by bare skin which adds oils. All of these raise the wet-bulb reading, increasing the measurement error so the operator overestimates the true humidity.

4.5 Pressure Sensors

Pressure is almost always measured as a differential pressure, either the difference between the pressures of two fluids or the difference in pressure between a fluid and a reference pressure. When the reference pressure is atmospheric pressure, we refer to the pressure of the fluid as gauge pressure. The name comes from the common use of pressure gauges that measure the difference in pressure between a fluid (such as water in a pipe or air in a duct) and the ambient air at the gauge location. The absolute pressure of a fluid is the gauge pressure plus atmospheric pressure (roughly 14.7 pounds per square inch at sea level).

Water pressure is typically measured in pounds per square inch, designated as psig (gauge pressure), psia (absolute pressure), or simply psi (differential pressure). Air pressures are generally measured in inches of water gauge, designated as inches H_2O, inches wg or sometimes wc (for water column). One inch H_2O is equal to 0.036 psi.

Mechanical Pressure Gauges: The Bourdon tube (see *Figure 4-22*) is the sensing element used in most pressure indicating gauges. It is a closed, spiral tube, connected at one end to the pressure being sensed, with atmospheric pressure as a reference. As the sensed pressure increases, the tube tends to straighten, and, through a linkage and gear, drives an indicating pointer. By adding a switch to the linkage (not shown), the device can become a sensor with switching capability.

A spiral tube is similar in principal to the Bourdon tube, but it is formed into a spiral spring shape that elongates or shortens as the sensed pressure changes.

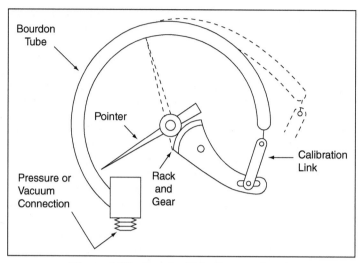

Figure 4-22   Bourdon Tube Pressure Sensor

The diaphragm sensor (see *Figure 4-23*) is an enclosure that includes two chambers separated by a flexible wall or diaphragm. The typical diaphragm is a thin steel sheet, sensitive to small pressure changes. Slack diaphragms of fabric are also sometimes used. Pressure differentials as low as a few hundredths of an inch of water gauge or as high as several hundred psi can be sensed (not by the same sensor, but sensors are classified by a wide range of pressure ratings). By means of appropriate linkages, the sensor output can also be used as a modulating controller or two-position switch. The latter is commonly used to indicate fan and pump status, proving flow indirectly by virtue of the fan's or pump's ability to generate a pressure difference.

These mechanical devices are made for a very wide range of pressures from fractions of an inch water gauge to thousands of pounds per square inch. Each gauge has a limited pressure range and, not surprisingly, the sensitivity reduces the greater the range. The Bourdon and spiral tube are typically used on water systems and diaphragm units on both water and air systems.

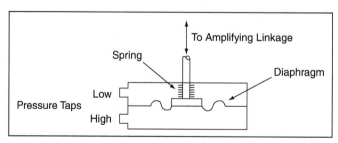

Figure 4-23   Diaphragm Pressure Sensor

These mechanical devices can all be connected to a transmitter. The transmitter detects the mechanical change and puts out a signal proportional to pressure. The signal may be electric or a change in air pressure for pneumatic systems.

You learned about resistance, capacitance, and inductance in Chapter 2. These three phenomena are used as the basis of the transducer constructed to measure electrical output from pressure sensors. The first, and simplest, is the potentiometer. This consists of a coil of resistance wire and a slider. As the mechanical sensor moves it moves the slider along the resistance coil. The change in position on the coil is sensed by the transmitter and converted to an output indicating pressure. The potentiometric unit is inexpensive and produces a high output but has low accuracy and extensive movement shortens the life.

The second type, the capacitance sensor is shown in *Figure 4-24*. The capacitance between two parallel charged surfaces changes as they move toward and away from each other. Again, a relatively inexpensive sensor but converting the signal to directly relate to pressure is not simple or inexpensive.

The inductive sensor is much like a transformer being two coils of wire around a metal core, *Figure 4-25*. The metal core is connected to the mechanical movement. As the core moves, the magnetic flux between the two coils changes and is measured by the transmitter. These units are both rugged and durable but, like the capacitance unit, converting the signal to directly relate to pressure is not simple or inexpensive.

Electrical Pressure Guages: These gauges all use an electrical method of detecting property changes. The first is one we have met before in temperature sensors, change in resistance. The strain gauge is a metal foil which changes resistance when stretched. The semiconductor version of the strain gauge has a higher output and is called the piezoresistive effect. These devices are bonded to a frame designed to distort under pressure. The transmitter detects the resistance change and converts it to a pressure signal output. The output from these sensors is often nonlinear, suffer from hysteresis (different reading on increasing pressure from decreasing pressure), and for accuracy need to be temperature compensated so sophisticated electronic circuitry. The resulting signal is proportional to differential pressure and may be used as a signal to an analog

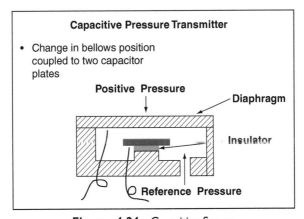

Figures 4-24 Capacitive Sensor

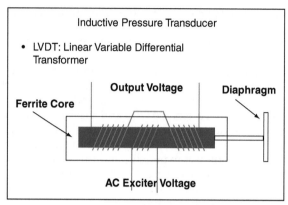

Figures 4-25   Inductive Sensor

electronic or digital controller. These devices are typically used on high-pressure water systems.

All the above sensors are designed to measure continuous pressures and changing pressures. The piezoelectric pressure sensor is different. In these devices the charge generated in a crystal under changing strain is detected, measured and amplified into a useful output signal. They only measure changing pressures, which they do very fast, with very low forces over a very wide range of strain. They thus find their niche in vibration and sound-sensing equipment.

4.6 Flow Sensors and Meters

The most common uses of flow sensors in air and hydronic systems are for energy process control and energy monitoring (sensors with indication and/or recording device called meters).

Typical processes using flow control include:

- Measuring the variable flow in large, chilled water plants to facilitate making decisions about flows and what equipment (typically chillers, pumps, and cooling towers) should be running.
- Flow measurement to adjust the flow through variable volume boxes.
- Using flow to adjust variable speed fans to maintain the correct flow balance between supply, return, and relief.
- Adjusting air flow through fume hoods to maintain the capture velocity under changing conditions.
- Indirectly assessing room-to-room pressure by the flow through specially shaped orifices.

Flow monitoring is used in both air and water systems where it is important to confirm flow before an action is taken, or to shut down plant on flow failure.

Energy monitoring is often done in chilled and hot water systems in order to assess energy costs. In multi-tenant buildings, this enables the landlord to apportion costs between the tenants.

Flow sensors commonly used in HVAC applications can be grouped into four basic categories:

1. Differential pressure flow sensors
2. Displacement flow sensors
3. Passive flow sensors
4. Mass flow sensors.

All types of flow sensors will only be accurate if the fluid flow is relatively fully developed and free of eddy and vortices caused by fittings and obstructions. Almost all sensors will require long runs of straight pipe or ductwork, on the order of 2–10 duct/pipe diameters upstream and about 2–3 diameters downstream, to provide an accurate signal. Where adequate distance is not available, straightening vanes or grids can often be used to improve accuracy. Some manufacturers have developed corrections for sensors mounted upstream of common obstructions (such as elbows in pipes), but still at some loss in accuracy.

The main variety of meters is in metering water flow. Although some meter types are common in water and air, we are going to start by considering water flow meters and then turn to air flow meters.

Differential Pressure Flow Meters: Correlating differential pressure to flow is one of the oldest techniques for flow measurement. Meters using this technique are all based on a form of Bernoulli's equation:

$$V = C\sqrt{\frac{2\Delta P}{\rho}}$$

(Equation 4-1)

where V is the velocity, C is a constant that is a function of the physical design of the meter, ΔP is the measured pressure drop, and ρ is the fluid density. Fluid density in both air and water systems is typically relatively constant over the normal range of operating conditions in HVAC systems. Some practical examples using this formula will come later.

Figure 4-26 shows an orifice plate meter. It consists of a plate with a round sharp edged shaped hole in the middle. If the Reynolds Number of the fluid through the plate is sufficiently high, the flow rate and pressure drop follow *Equation 4-1* very closely (their accuracy becomes erratic and unpredictable at low flows). One of the unique aspects of orifice meters is that the flow coefficient C can be determined from basic principles using the measured area of the pipe and orifice opening. For most other devices, the flow coefficient must be determined by experimentation using some other, more accurate device. For this reason, orifice meters are often used to calibrate other meters. They are not commonly used in HVAC systems because the pressure drop across them is high relative to other meters. Their accuracy is poor at low Reynold's numbers and it degrades relatively quickly because the orifice is prone to dirt accumulation and wear around the opening due to the abrupt velocity change at that point.

Figure 4-27 shows another flow meter, the Venturi type, based on the same principle but the pressure drop is reduced because of the smooth inlet and outlet. As the fluid is accelerated, static pressure is converted into an increase in velocity (kinetic energy). By measuring the drop in static pressure from the

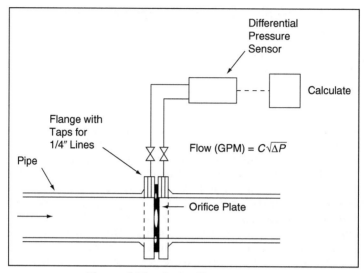

Figure 4-26   Orifice Plate Flow Meter

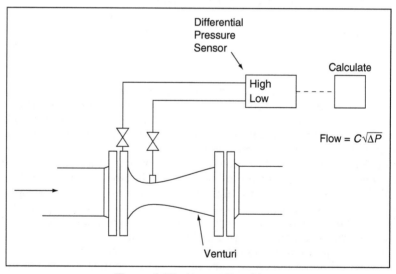

Figure 4-27   Venturi Flow Meter

inlet to the most constricted part of the meter, the velocity can be determined using *Equation 4-1* and a flow coefficient that is generally determined from bench tests.

Venturi meters are commonly used for steam flow measurement. They are less commonly used for water flow measurement and almost never used for air flow measurement because alternative technologies are less expensive.

For water flow at standard density, the pitot tube equation can be written:

$$V = 12.2\sqrt{\Delta P}$$ (*Equation 4-2*)

where ΔP is in psi and V is in feet per second (fps). *Figure 4-28* shows a pitot flow sensor as commonly configured in a pipe. Velocity pressure sensing ports are located at specific points along the tube to compensate for the natural distortion of the velocity profile in the pipe. Some inaccuracy can occur due to turbulence caused by the sensor itself, which affects the downstream static pressure reading.

Some manufacturers use a shape other than round for the probe to improve accuracy. An example is shown in *Figure 4-29* of a commonly used sensor:

The shape of the bar improves the accuracy compared to a round pitot tube ($\pm 1\%$ versus $\pm 5\%$), particularly at very high or very low velocities. Annubar® sensors are bi-directional; they can measure flow in either direction because the two sensing ports are symmetrical. A bi-directional pressure transmitter must also be used in this case (bi-directional means that it will read on both sides of the zero mark, positively and negatively at times). Bi-directional sensing is not a common requirement of HVAC systems, but it can be used in the common (decoupling) leg of a primary-secondary piping system for chiller staging control.

The accuracy of differential pressure flow sensors will vary strongly as a function of the transmitter that is used to convert the signal to either a

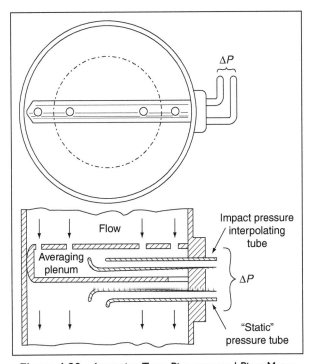

Figure 4-28   Averaging Type, Pipe-mounted Pitot Meter

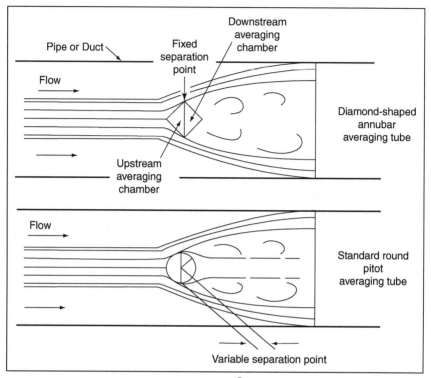

Figure 4-29   Annubar® Flow Sensor

pneumatic or electronic signal for use by the control system. As the pressure signal falls, the accuracy of the flow sensor also falls. The transmitter must be selected for the maximum pressure (maximum velocity) it must sense, but this usually makes it oversized for low flows and accuracy falls off quickly at velocities below about 25% of the maximum.

Displacement Flow Meters: Displacement flow meters work by using the fluid to rotate or displace a device inserted into the fluid stream. A simple example is the paddle flow switch shown in *Figure 4-30* which is commonly used to "prove" fluid flow in a pipe. It is screwed into a tee or weldolet and includes a switch that is activated when fluid flow deflects the paddle. A similar device with a larger and lighter paddle called a sail switch is used in air systems. Both of these devices are sensitive to physical damage, dirt, and corrosion, and require regular maintenance to ensure reliability.

Figure 4-31 shows a turbine meter that measures the flow by counting the rotations of a propeller-shaped rotor placed in the fluid stream. This is a very common sensor for water flow measurement in HVAC systems. Rotations are commonly counted using magnetic sensors (requiring a metal rotor), infrared light reflections from the blade being counted, or nonmagnetic radio frequency impedance sensors (requiring some electrical conductivity of the fluid; fluids other than deionized water). Some turbine meters are bidirectional. Accuracy can be improved by using a dual turbine meter that has two turbines mounted in parallel with rotors that rotate in opposite directions. These meters offer improved accuracy

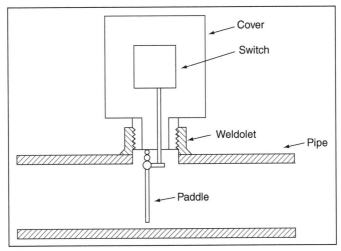

Figure 4-30   Paddle Flow Switch

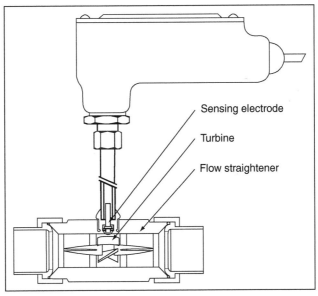

Figure 4-31   Inline Turbine Flow Meter

because they cover more of the flow passage, and the counter-rotating rotors can cancel out swirling currents caused by upstream or downstream elbows.

A tangential paddlewheel meter (not shown) is similar to a turbine meter, with the rotor replaced by a paddlewheel mounted tangential to the flow. Flow rate is also determined by counting rotations. The advantage of these meters is that they are relatively reasonable priced and accurate, when used appropriately. Sometimes they can have multiple paddles connected in tandem for more accuracy. The disadvantage of this meter compared to a turbine meter is that the tangential orientation increases wear on the bearings, requiring more frequent replacement of the rotating element, but, all in all, the paddle-type flow meters are one of the best values in the flow sensor industry.

A target meter (see *Figure 4-32*), also called a drag-force meter, measures flow rate by the amount of stress in the stem supporting a paddle or other obstruction mounted in the flow stream. The higher the flow rate, the greater the bending action, and the greater the stress. Stress is typically measured using a strain-gauge located where the support stem is attached to the meter body. One advantage of this device is that it has no moving parts.

A vortex meter measures flow rate by electronically measuring the pattern of flow around a shaped sensor inserted in the pipe. This device is very sensitive and accurate, but it is not commonly used in HVAC work, because it is very expensive.

Other displacement meters measure flow by timing how long it takes for the fluid to fill up a container of a known volume. Unlike most other sensors, these sensors are accurate for low flow rates because of our ability to measure time accurately. This type of sensor is used primarily to calibrate other sensors.

Passive Flow Meters [14]: Passive flow meters measure flow without placing any obstructions in the fluid stream. Therefore, they create no additional pressure drops and have no moving parts in the fluid stream that require maintenance.

Transit time ultrasonic meters measure flow rate by detecting small differences in the time for sound waves to move through the fluid as fluid velocity varies. As shown in *Figure 4-33*, ultrasonic sound waves are shot at an angle through the fluid and detected by a sensor downstream. The sensors may be clamped onto the outside of the pipe without having to penetrate the pipe at all. The reported accuracy of transit-time sensors is ±1% of full range with

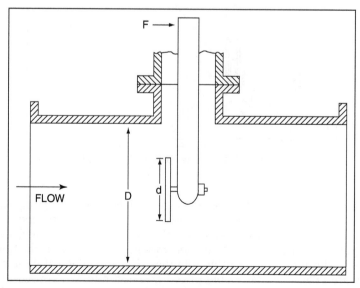

Figure 4-32   Target Meter

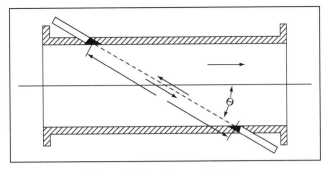

Figure 4-33   Ultrasonic Meter

fully developed very clean water, no air bubbles, and straight flow. However, no more than ±5% should be expected in practice because of the variability in piping dimensions, fluid properties, and other practical limitations.

Another type of ultrasonic meter is the Doppler effect meter. It measures the Doppler shift in the frequency of the sound waves caused by the fluid flow. Accuracy is somewhat less than transit-time sensors and the fluid must have a certain amount of impurities to deflect the signal. The Doppler effect sensor provides 5%–10% accuracy in actual practice. Its advantage over the transit time sensor is that it works regardless of water quality, so it can be used in sewage plants and irrigation systems.

Magnetic flow meters measure flow rate by magnetic induction caused by the moving fluid when exposed to a strong magnetic field. The fluid must have a non-zero electrical conductance (not deionized water). These expensive sensors are very accurate and can be used for a wide range of fluids including sludges and slurries, and are relatively insensitive to turbulence.

Mass Flow Meters [14]: While mass flow meters are available for liquid flow sensing (such as Coriolis force meters, angular momentum meters), they are not commonly used in HVAC applications.

Having covered water flow meters let us turn to the measurement of air flow.

Mass flow meters do exist for water systems but are not used in air systems the opposite is true for air systems. One type of mass flow meter is very popular in air flow measurement: the thermal or hot-wire anemometer.

The thermal anemometer (see *Figure 4-34*) uses a heated probe placed in the air stream that is cooled by the movement of air in direct proportion to its mass flow rate. The probe consists of a temperature sensor and electric resistance-heating element. The device measures the electrical current required to keep the element at a constant temperature (around 200°F) and translates this into a velocity signal (assuming a constant air density) that can be read on a meter or used in a control system.

In modern thermal anemometers, the heating element and temperature sensor are often the same device, usually a self-heated thermistor. Often, another temperature sensor is installed upstream of the probe to measure entering air temperature, used in the electronics to determine air density to provide a more accurate velocity signal. These anemometers are called temperature compensated. The temperature sensor usually can be used by the control system for other purposes as well, often obviating the need to add another air temperature sensor.

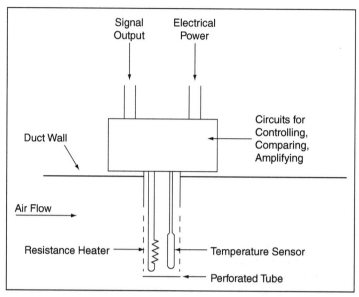

Figure 4-34   Thermal Anemometer

Thermal anemometers have the advantage of being able to sense much lower air velocities than pitot tube sensors. Common commercial models maintain ±2% to ±3% accuracy down to below 500 fpm; below this, accuracy drops to about 10–20 fpm. A low limit of about 100 fpm can be sensed with an accuracy of ±20 fpm. By comparison, pitot sensors are seldom accurate below about 400 fpm (0.01 inch wg) or even higher, depending on the transmitter used. Moreover, the pitot sensor will not have the broad range of the thermal anemometer, which can read velocities from about 150 to 5000 fpm with reasonable accuracy. Pitot sensors are limited in range by the transmitter, which will have a range of no more than about 3 or 4 to 1 while maintaining at least 10% accuracy. Therefore, thermal anemometers are a better sensor for some VAV systems that might experience a wide air flow operating range, and for outdoor air intakes on economizer systems where a wide range is typical. A relatively new product is a thermal anemometer designed to be mounted in the fan inlet, a preferred location as described above for pitot sensors.

Thermal anemometers can also be used to measure differential pressure across a barrier, such as a wall between two rooms. Special mountings are used with a very small porthole through the wall to allow air to pass. This velocity of this air flow is measured and can be translated into a pressure difference by means of the *Equation 4-3*. The accuracy of the anemometer in this application is more accurate than a differential pressure transmitter when the pressure difference being measured is very small, less than about 0.02 inch wg.

An example of an air flow meter is shown in *Figure 4-35*.

Pitot tube sensors (see *Figure 4-36*) are commonly used to measure the speed of aircrafts. They are also used very commonly for measuring both air and water flow in HVAC applications. The differential pressure in *Equations 4-1–4-3* in this case is the difference between the total and static pressures, a quantity called the velocity pressure. As can be seen in *Figure 4-36*, the inner tube senses

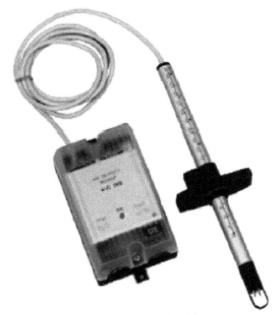

Figure 4-35   Air Flow Meter

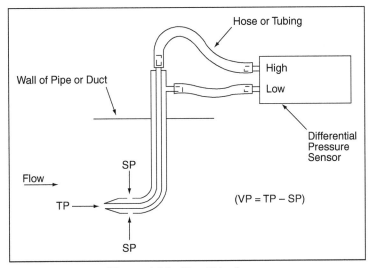

Figure 4-36   Pitot Tube Sensor

total pressure of the fluid, which due to the static pressure plus the force exerted by the fluid's velocity, called the velocity pressure. The outer tube has openings in the sides, which are not impacted by the fluid flow and therefore sense only static pressure. The difference between the two is used in *Equations 4-1–4-3* to determine fluid velocity. In an ideal pitot tube, the C coefficient is a constant regardless of geometry.

For air flow at standard density, velocity may be calculated from the pitot tube velocity differential pressure as:

$$V = 4005\sqrt{\Delta P}$$ (*Equation 4-3*)

where ΔP is measured in inches of water gauge (wg) and V is measured in feet per minute (fpm).

Figure 4-37 shows an air flow measuring station (FMS) commonly used in duct applications.

In large ducts, the FMS is composed of an array of pitot sampling tubes the pressure signals of which are averaged. This signal is fed to a square root extractor, which is a transmitter that converts the differential pressure signal into a velocity signal. (With digital control systems, this calculation can be made in software with improved accuracy over the use of a square root extractor.) The velocity measured in this way is an approximate average of duct velocity, but it is not a precise average because pressure and velocity are not proportional (the average of the square of velocity is not equal to the average of the velocity).

Where fan air flow rate measurement is required, the preferred location of the pitot sensor is in the inlet of the fan. Two arrangements are available. The first, which can be used with almost any fan, has two bars with multiple velocity pressure and static pressure ports mounted on either side of the fan axis. The second, much less common arrangement has multiple pinhole pressure taps that are built into the fan inlet by the fan manufacturer. Differential pressure is measured from the outer point of the inlet to the most constricted point, much like a Venturi meter.

Locating the air flow sensor in the fan inlet has many advantages compared to a duct mounted pitot array. First, air flow is generally stable in the inlet (except when inlet vanes are used, in which case this location is not recommended) and

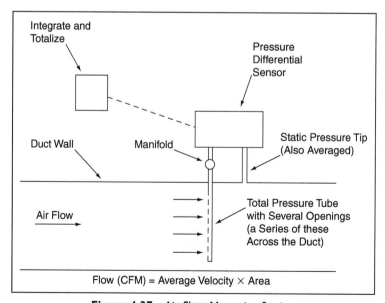

Figure 4-37   Air Flow Measuring Station

velocities are high, which increases accuracy because the differential pressure signal will be high. Even where inlet vanes are used, a location in the mixed air plenum space could be found. This location also reduces costs because the sensor array is smaller than the array required in a duct. Perhaps the most important advantage of this location is that it obviates the need to provide long straight duct sections required for the duct mounted array. The space needed for these duct sections seldom seems to be available in modern HVAC applications where the operating space occupied by HVAC systems is heavily scrutinized by the owner and architect, reduced to its smallest area possible, and consciously minimized.

Displacement Flow Meters [14]: Displacement flow meters work by using the fluid to rotate or displace a device inserted into the fluid stream. A simple example is the paddle flow switch which you saw in *Figure 4-29*. A larger and lighter paddle called a sail is used in air systems.

Propeller or Rotating Vane Anemometers: Propeller or rotating vane anemometers are commonly used hand-held devices for measuring air velocity. They are seldom used as sensors for control systems because they are accurate only for measuring velocity at a single point, and velocity in a typical duct system varies considerably over the duct face.

Table 4-6 summarizes some of the velocity and flow sensors that we have been discussing.

Table 4-6   Table of Flow Sensors (Hegberg, 2001–2002)

Velocity and Flow Sensor Summary				
Sensor Type	Primary Use	Accuracy and Maximum Range	Advantage	Disadvantage
Orifice plate	Water	±1–5%, 5:1	Inexpensive, great selection	Sharp edge can erode lowering accuracy
Venturi	Water, high-velocity air		Low head loss	Expensive, large
Turbine	Water	±0.15–0.5%, up to 50:1		Blades susceptible to damage
Ultrasonic	Water	0.25–2%, 100:1		
Vortex shedding	Water	±0.5–1.5%, 25:1		
Pitot tube	Air flow	Minimum velocity 400 fpm	Inexpensive	Can plug with dirt, limited in lowest velocity
Thermal anemometer	Air flow	±20 fpm at 100 fpm	Good at low velocities, small sensor easy to insert into duct	Dirt can reduce accuracy
Rotating vane	Hand-held air flow		Inexpensive	Not robust, large

4.7 Auxiliary Devices

In addition to controllers and sensors, most control systems will require additional devices to completely implement the desired control sequence. This is true whether the control system is electric, analog electronic, or digital. Many of these auxiliary devices that are commonly used with all control system types are discussed in this section. These devices are absolutely needed for every control system in order to make it work per its sequence. Devices that are specific to the primary control types are discussed in the following chapters.

[C]*Relays:* A relay is a device for amplifying, varying, or isolating a signal as shown in *Figure 4-38*. This includes changing the signal type (from electrical to pneumatic), the amplitude (from low to high voltage or the reverse), and time delay or signal reversal. Isolation means that the control signal is electrically separated from the controlled circuit.

Electromechanical relays were introduced in Chapter 2. While they are electrical devices, they are commonly used in pneumatic, electronic, and digital control systems. In the case of digital controls, relays are primarily used to start and stop equipment when the controller contact has an insufficient current rating to power the equipment or starter directly.

Relay contact capacities are rated in amperes, inductive, or resistive. Inductive ratings are used for power to loads such as motors, transformers, solenoids, etc., because breaking the power contact creates a back-emf (electromotive force) that causes an arc to form as the contact opens. The contact material must be resistant to damage from such arcs. When used for incandescent lighting or other resistive loads, the back-emf is small, arcing is not such a serious problem, and ampere ratings can be higher.

Solenoid coils are available in a wide range of voltages from a few volts dc in electronic work to 480 V ac or more in power relays. The voltage being switched by the contact is, more often than not, different from the voltage on

PAM-1

Figure 4-38   Relays (Kele, 2002)

the solenoid. Be careful when switching solenoids that the proper VA required is being provided by the transformer.

Relays are loosely divided into power or control relay classifications. This is arbitrary as a function of usage, contact ratings, and voltage and power requirements. They are further classified as electrically held or latching type. Electrically held means that a spring returns the relay to the normal position when the power to the solenoid is removed. Latched means that power to a solenoid is applied momentarily to drive the relay in one direction, where it stays until power to a second solenoid causes it to drive to the other position. This is useful when the relay needs to be maintained in the energized position even when control power fails.

Another type of relay is the timing relay, used for time delay or timed programs. Timing relay contact arrangements include:

- Normally open, instant open, time delay close upon energization. This means that when the relay is energized, the contact closes only after a time delay. The contact opens immediately (instant open) upon de-energization of the relay.
- Normally closed, instant close, time delay open upon energization. This is similar to the previous relay except that normally closed contacts are used.
- Normally open, instant close on energization, time delay open after de-energization. With this relay, the contact will close right when the relay is energized and remain closed for some time after power is removed from the relay coil.
- Normally closed, instant open on energization, time delay close after de-energization. This is similar to the previous relay except that normally closed contacts are used.

See Chapter 2, *Figure 2-32* for symbols commonly used to represent these different relay types.

Time delay relays are very useful in implementing control logic with electric, pneumatic, and analog electronic control systems. With DDC (direct digital controls), time delays are typically effected in software.

Today, most timing relays are electronic, but in the past they were pneumatic. Some electronic relays that provide a time delay on energization can be wired in a series fashion with the load, rather than in a parallel fashion. Both styles are shown in *Figure 4-39*. Normally, powered devices cannot be wired in series (see Chapter 2), but with this relay, the electronics in the relay that provide the time delay work off a very low current (high resistance). This current is so low that the voltage passes to the load when the relay is energized but before the timer has timed-out it is insufficient to cause the load to be energized, and the current is so low that the resistance of the load is insufficient to cause a large voltage drop across it. Once the timer has timed-out, a contact is closed and power is passed to the load as with the standard parallel-wired relay.

Fan and Pump Status Switches: Fan and motor operating status are often important to the proper operation of the HVAC system; a current switch is shown in *Figure 4-40* (Kele, 2002).

When a pump or fan fails to operate, it might be critical to interlock other equipment (such as a direct expansion air-conditioner or a water chiller) to prevent their operation. In its simplest form, status is taken from an auxiliary contact in the motor starter (Chapter 2), or a relay in parallel with the starter

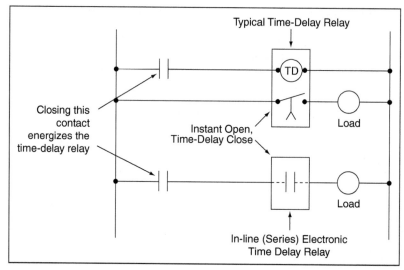

Figure 4-39   Time-Delay Relay Wiring

Figure 4-40   Current Switch

coil or motor. However, this is only an indirect indication that the fan or pump is on. If the motor fails, the power disconnect at the motor is opened, or if fan belts or pump couplings fail, a contact wired in this fashion will still indicate normal operation.

For this reason, it is desirable to sense the real operating condition of the fan or pump or the motor driving it. In common practice, this is done by means of a paddle or sail switch in the duct or piping, a differential pressure switch across the fan or pump, or a current switch mounted on the power wiring to the motor.

The flow and differential pressure switches (discussed in the sensor sections and an example shown in *Figure 4-41* (Kele, 2002) tend to be problematic in practice, particularly on fan systems due to sensitive set points and failures of paddles over time, however they are still widely used to sense filter status, and air flow proving for electric heaters. The current switch is rapidly taking their place because it is more reliable and less expensive to install. This device includes an induction coil as part of a bridge circuit, which allows it to sense the current flow in one phase of the power wiring. The sensed value is compared to a set point that corresponds to a normal operating current. The set point can be high enough to distinguish between an unloaded motor (as might occur if a belt or linkage was broken) and a blocked tight condition (as might occur if a damper or valve had shut off, completely blocking flow). Therefore, the current relay is almost as dependable at indicating true fan or pump status as a flow switch, more dependable than a differential pressure switch (which is not able to distinguish between the blocked-tight condition and normal operation), and it is significantly more reliable and less expensive to install.

Timeclocks: Timeclocks are used for starting and stopping equipment on a regular cycle. Mechanical timeclocks are available for 24-h or seven-day programs. On/off initiation is generally accomplished using pins attached to a wheel that rotates with time. The pin's location on the wheel determines the start and stop times. When the pins rotate past a switch, they push it either open or closed. Mechanical timeclocks generally need to be adjusted for time changes (for example, standard to daylight savings) and must be reset after power failure. Battery backup is sometimes provided to avoid the latter problem.

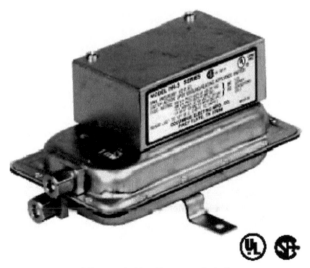

Figure 4-41   Pressure Switch

Electronic timeclocks, also referred to as programmable timeclocks, have largely taken the place of mechanical timeclocks, offering the same features and more. Scheduling capability varies, but typically they can control equipment on the basis of two day-types per week (for example, weekend, weekday) or on a seven-day basis, with one to four on-off periods each day. Some are capable of 365-day programming. Typically, on/off schedules are programmed through a keypad on the timeclock face. Most have capacitor or battery backup to retain programming in case of power failure.

Wind-up Timers: A manual wind-up time switch is a spring return switch that can be turned on manually to provide a given on-time for the controlled device as desired by the operator. Switches are available for times of a minute or less, or up to several hours. These are often used to temporarily bypass timeclocks to allow use of air-conditioning equipment, for example, on irregular schedules while providing for automatic shut-off when no longer needed.

Limit Switches: Limit switches are used in many control applications. The basic switch is single or double throw, single or double pole. When activated by a plunger or lever (see *Figure 4-42*), it is used for position sensing, as with a damper or valve. It can also be activated by pressure (typically through a bellows) for use in refrigeration cycles or with steam or hot water boilers.

Manual Switches: Manual switches are available in many configurations such as lever, rotary, and pushbutton arrangements. Lever switches are usually single or double throw, or single or double pole. (For a description of the terms *throw* and *pole*, see Section 2.4.) Rotary switches are used for selecting one of three or four operating modes. Pushbuttons are either momentary contact or maintained contact. In the maintained contact style, they perform like a rotary switch. In the momentary contact arrangement, the contact remains closed only while the button is manually depressed; they require an external sealing or maintaining contact.

High and Low Limit Switches: High and low temperature, level and pressure alarm, and safety switches are needed in steam and water heating systems, as well as refrigeration systems. These are the same as standard control sensors, but they are connected to provide an alarm and/or shut down the HVAC system. Examples include high discharge pressure on refrigeration compressors, off-normal steam pressures, low temperature for coils-freezestat

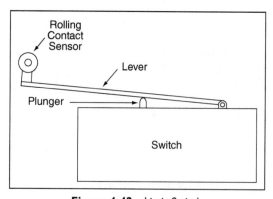

Figure 4-42   Limit Switch

(*Figure 4-43*), off-normal hot water temperatures, sudden pressure drops in high temperature water systems and low water level in cooling tower sumps.

A special low limit switch is the freeze-stat, which is used in HVAC air system mixed-air or outdoor air streams to prevent freezing temperatures from reaching water or steam coils. The device is a bulb and capillary sensor (see *Figure 4-44*) with a long bulb that is filled with a refrigerant. Where stratification is expected (the usual condition at an outdoor air intake), the bulb must

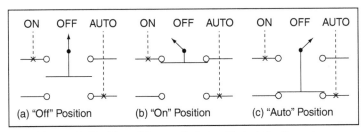

Figure 4-43   Manual Rotary Switch – Typical Configurations

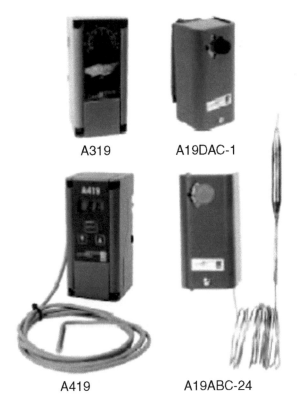

A319   A19DAC-1

A419   A19ABC-24

Figure 4-44   Low/High Limits (Typical for Cooling coils) (Kele, Johnson Controls, 2002)

be long enough to form a grid across the duct so that no portion of the coil is left unprotected. The refrigerant is selected to be a gas at temperatures above about 35°F. If any small segment of the bulb senses a lower temperature, the refrigerant in that segment will condense, causing a sharp pressure drop in the bulb, which is used to trip a switch in the device. Normally, this switch is wired to shut down the fan and close the outside air intake.

Fire and Smoke Detectors: Fire or smoke detection instruments are required in most air handling systems by Code. Fire-stats (switches sensitive to high temperatures) are used mostly in small systems (under 2000 cfm), while some codes generally require smoke detectors in larger systems at main return air and/or supply air ducts. The exact position required for the detector varies with different building codes. Some codes require that detectors be placed in the return air stream before air is mixed with outdoor air to sense smoke being generated in the spaces served by the system. Others call for the detector to be placed downstream of the fan and filter to prevent recirculation of smoke from almost any source (such as from a filter fire or from the outdoors).

A duct-type smoke detector is shown in *Figure 4-45*. Note the sampling tubes required in the duct. These are arranged to extract an air sample from the duct through the longer, upstream tube and then return it through the shorter tube. The upstream sampling tube has ports much like a pitot type flow sensor. To function, a minimum velocity must be maintained, generally at least 300–400 fpm and preferably higher.

Smoke detectors are available using primarily two detection technologies: photoelectric and ionization. The photoelectric detector is sensitive to larger particles but not to fine particles (less than about 0.1 µm). The ionization detector is sensitive to these fine particles but less sensitive to large particles. Because most air filters are efficient in capturing large particles, the ionization detector is the most appropriate type for use in supply duct systems downstream of filters. As smoke moves away from the fire source, smoke particles tend to agglomerate into larger particles. Therefore, the photoelectric detector

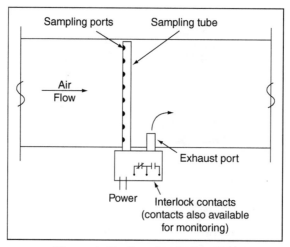

Figure 4-45   Duct Smoke Detector

may be the more appropriate type to use for return or exhaust duct installations, although both styles are common.

In most cases, smoke detectors and fire-stats simply shut down the air system either directly, by hard-wiring the detector to the fan start circuit (see Chapter 2), or indirectly, by signaling the detection of smoke to a fire alarm system which, in turn, initiates fan shut-down through fire alarm remote controlled relays (see Chapter 2). If the air system is designed as a smoke control system, then the smoke detector might initiate a smoke control sequence; the exact sequence will vary depending on the design of the system. Smoke control performed as a part of the HVAC control systems is very cost effective and efficient. Smoke detectors are typically furnished as a part of the fire alarm system, but sometimes we are able to monitor its status through auxiliary contacts in the detector.

Indicating Lights: Pilot lights come in a wide range of sizes, types, and colors. In HVAC systems, they are used to indicate the on or off condition of motors (fans, pumps, etc.) as well as normal and alarm conditions.

Audible Alarms: In addition to a visible light, an audible alarm such as a horn or bell is sometimes desired. They are usually provided with a silence switch, as shown in *Figure 4-46*. The wiring in this diagram allows the horn to be silenced while the light continues to indicate an alarm condition until the alarm sensor is cleared.

Monitoring: Monitoring instrumentation is needed to provide information on operating conditions and the need for maintenance. Temperature, humidity, and flow indication may be local or remote. Remote indication implies a central monitoring panel or a computer. Local temperature indicators include mercury, liquid, and bimetal thermometers. Remote indication requires the use of transmitters, either electrical or pneumatic. Local humidity indication is not normally used but remote indication is feasible, using electronic sensors and transmitters. Local or remote mounted flow indicators, such as manometers that use a column of water with lines calibrated to indicate air flow, may be based on the same flow sensors used for control.

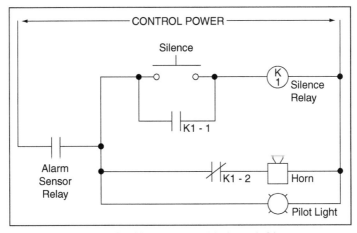

Figure 4-46 Alarm Horn and Light with Silencer

Figure 4-47   Vibration Sensor (Kele, 2002)

Isolation Room Pressure Sensing: For hospital and clean room applications, room pressurization sensing and display is used more frequently as the benefit–cost ratios are increasing. The relative pressure between the patient room and the adjacent hallway is measured, and visual and audible alarms are generated when the "negative" or "positive" set points are exceeded.

Vibration Sensing: The sensor in *Figure 4-48*, used for equipment that moves, rotates, spins, and has internal motion, a vibration sensor is an excellent way to sense when something is wrong and provide an alarm signal. Typically, the vibration sensor causes the control system to turn off the equipment that is vibrating in order to limit damage. These are commonly used for cooling tower fans.

CO_2 and Indoor Air Quality (IAQ) sensing: The IAQ sensors shown in *Figure 4-48* are used all over buildings today to sense the conditions and alarm the occupants so that corrective action can be taken. For example, CO_2 can be used as a surrogate to sense excess bioeffluents in the air, and therefore causing an exhaust fan and/or a ventilation damper to open. IAQ sensors can be made to sense a multitude of gases such as carbon monoxide, refrigerants, ammonia, acetylene, nitrous oxides, sulfur dioxides, chlorines, Hydrogen Cyanides, methane, natural gases methane, propane, hydrogen chloride, etc. Typically, the outputs of these sensors are an industry standard 4-20 ma or 0-10 vdc, so that they may be used with a multitude of controllers.

Multiplexers: These devices (*Figure 4-49*) take an output, such as a pulse width signal, and analog 4–20 ma/0–10 vdc , and convert it to a staged relay

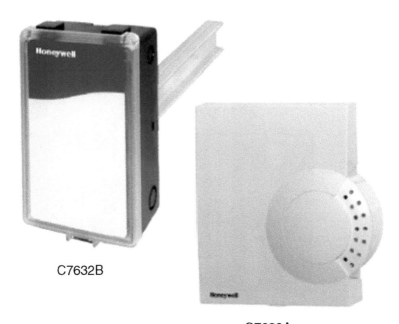

C7632B

C7632A

Figure 4-48 CO$_2$ and IAQ (Kele, Honeywell, 2002)

Output Multipexer

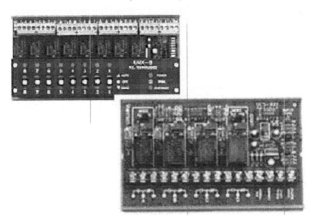

Figure 4-49 Multiplexers (Kele, 2002)

output that drives several individual relays on the multiplexer board. Typically, a modulating control signal can stage on and off several individual points this way. For example, if you wanted to take a modulating analog signal and stage on and off six stages of electric heat, then you could feed the multiplexer with the modulating signal and have it stage the heater.

Shown with Optional Pressure Gauge

Figure 4-50   Transducer with Optional Pressure Gauge (Kele, 2002)

Electric and Pneumatic Transducers: These devices (*Figure 4-50*) are extremely useful in the day-to-day activities of making control systems work under all kinds of different signal conversions. They take a modulating output and change it from electric to pneumatic modulating, in proportion, or vice-versa. For example, an electric to pneumatic transducer, or EPT, will take a 4–20 ma signal and convert it to a corresponding 0–20 psig signal for driving pneumatic devices.

Similar to these devices are their two-position devices, called an Electric to Pneumatic relay (EP), and a Pneumatic to Electric relay (PE). These control devices are two-position and snap-acting.

Fluid Level Devices: Fluid level devices (*Figure 4-51*) sense the fluid levels of their containment area and either send an analog or digital signal output. For example, in a drain pan of an air-handling unit, a drain pan float switch, which senses its fluid level, opens a contact and stops the unit and/or valve and/or cooling effect. Again, a typical cooling tower basin has a fluid level device, sometimes analog and sometimes digital, that senses the water level in the basin, and, when it is low, opens a valve to let in make-up water. Another interesting fluid level sensing is done in ice tanks, where the fluid level rises as the ice freezes, and finally signals a time for the ice-making to stop for that batch.

DDC systems can provide timely and cost effective monitoring and alarms via remote telephonic and internet connection to websites, pagers, printers, cell phones, call stations, etc. (Hydeman, 2002).

Figure 4-51   Fluid Level Device

The Next Step

In the next chapter, we will learn about self-powered and system powered controls. These are controls that, as their name suggests, obtain their motive power from internal property changes such as expansion or from the system such as velocity–pressure in air systems. Although limited in performance, these controls have the advantage of not requiring an electrical supply which can provide adequate performance at an affordable cost and a very high level of reliability.

Bibliography

ASHRAE (1991) *Terminology of Heating, Ventilating, Air Conditioning, and Refrigeration.* Atlanta, GA.

ASHRAE (2001) *Design and Specification of DDC Systems.* Atlanta, GA: ASHRAE Learning Institute.

Harriman, L., Brundrett, G. and Kittler, R. (2001) *Humidity Control Design Guide for Commercial and Institutional Buildings.* Atlanta, GA: ASHRAE.

Hegberg, R. (2001–2002) *Various Presentations from the Little Red Schoolhouse.* Chicago, IL: Bell and Gossett.

Hydeman, M. (2002) *Comments on SDL Controls PDS.* Alameda, CA. Taylor Engineering.

Kele (2002) Solutions Catalog available on-line: www.kele.com.

Chapter 5

Self- and System-powered Controls

Contents of Chapter 5

Study Objectives of Chapter 5

Self-powered and system-powered controls are those that do not require an external power source such as electricity or pneumatic control air. This chapter explains how these devices work and where they are commonly used.
   After studying this chapter, you should be able to understand:

   The sources from which self-powered controls derive their control power.
   How a thermopile works.
   How air flow can be modulated using self-powered variable volume diffusers.
   The advantages and disadvantages of system-powered controls.

5.1 Principles of Operation – Self-powered Controls

Self-powered controls are those that draw the energy needed for their operation from the systems that they control or operate. They are commonly used on small systems or individual units where they are more convenient and less costly because they do not require an external power source such as electricity or pneumatic control air.
   The power source in self-powered devices is typically derived from:

- Electrical potential generated by a thermopile, which is a type of thermocouple operating as a power source (like a battery). Like a thermocouple, the thermopile is composed of two dissimilar metals bonded together that create voltage. The magnitude of the voltage varies with the temperature to which the device is exposed. With thermocouples (Chapter 4), this effect is used to measure temperature. With the thermopile, the same effect is used as a source of power. Essentially, the thermopile converts thermal energy to electrical energy. The power and voltage are very small (measured in milli-watts and milli-volts) and the controlled devices used with them must be specifically designed for this application.

- Pressure resulting from expansion and contraction of a temperature-sensitive substance. The substance is selected to have a large coefficient of expansion; it will greatly expand when its temperature increases, thereby increasing the pressure on a diaphragm that in turn can drive a valve or damper. This is a power version of the bulb-and-capillary sensors that we covered in Chapter 4 (*Figure 4-5*).
- Pressure from the fluid being controlled. In the most common air system application, the pressure of the air stream being modulated is used to inflate a bellows that in turn moves a damper or bladder to control the air flow rate.

5.2 Examples of Self-powered Controls

Thermopile Controls

The thermopile is a widely used device in many gas-fired heating systems for residential or small commercial installations and in gas-fired unit heaters. The source of heat is the flame of the gas standing pilot light, as shown in *Figure 5-1*.

The electrical potential (voltage) generated by the heat of the flame provides the power for opening and closing the gas valve. A thermostat completes the circuit. The electric power generated by the thermopile is very small, so the gas valve is designed to utilize gas pressure to assist in opening the valve. Because the system will not function if the flame is not lit, the device is also a safety control. This also requires that the pilot light remains lit whenever the system must operate. Standing pilots waste energy and are prohibited in many situations by modern codes such as ANSI/INIESA/ASHRAE Standard 90 1-2004.

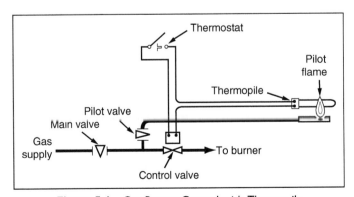

Figure 5-1   Gas Burner Control with Thermopile

The thermopile was a particularly convenient control for gravity furnaces, a now obsolete furnace design that used the buoyancy of the heated air as the motive force for distributing heat through the duct system instead of a fan. Because these furnaces had no fan, the use of a thermopile meant that no power source other than gas was needed.

Hot Water Control Valve

Another self-powered device that is still fairly common on small radiators and baseboard heaters is the control valve shown in *Figure 5-2*.

The valve has a bulb and capillary containing a fluid that expands and contracts to drive a bellows which in turn drives the valve stem. A spring is provided for return to the closed position. This system includes power source, sensor, controller, and controlled device in a single package. A somewhat simpler form of this device is used in hot water radiator valves where the sensor is incorporated into the valve.

Self-powered VAV Diffuser

Probably the most common self-powered device used in modern HVAC air systems is the self-powered variable air volume supply air diffuser. (Note that self-powered variable air volume supply air diffusers are significantly less common than electrically, or pneumatically powered units.) To control room temperature, the diffuser is designed so that as air is supplied through the diffuser, some room air is induced across the bulb of a bulb-and-capillary sensor/controller that is linked to a damper in the diffuser. The material used in the bulb is most commonly a type of wax that has a very high thermal expansion coefficient, and thus is able to generate significant torque to operate

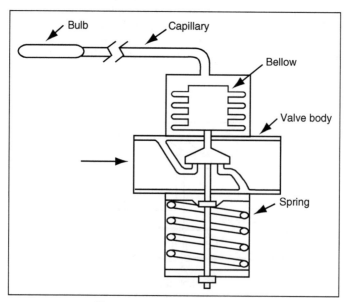

Figure 5-2   Self-Powered Valve

the damper. The operation of a self-powered VAV diffuser is similar to the valve operation represented in *Figure 5-2*. As the room temperature varies, the controller modulates the damper to control the air flow through the diffuser. The temperature range over which the controller set point can be adjusted is small; typically no greater than 70–80°F.

The maximum air pressure against which the self-powered VAV diffuser can operate is also small (about 0.25 inch wg), so the system must be carefully designed. On systems with long duct runs, the duct pressure behind outlets nearest to the fan may exceed the maximum rating. If so, the duct system must be broken into multiple branches, each with a duct mounted damper controlled to limit duct static pressure in the duct downstream (see *Figure 5-3*).

On small systems, duct pressure may be below the maximum when all outlets are open, but pressure can increase as the fan rides up its curve when air outlets close. Some means of static pressure control is usually required (such as variable speed drives or inlet vanes used on standard VAV systems). More commonly in small systems, pressure can be controlled by using a special bypass VAV diffuser that bypasses air to the ceiling (which must be used as a return plenum) when it is not supplied to the space. This maintains a relatively constant airflow rate that keeps the supply fan from riding up its curve and generating additional duct pressure. It also keeps a constant volume of air across the cooling/heating coil, an important consideration on small direct-expansion air conditioners and heat pumps.

One big advantage of self-powered variable air volume diffusers is that each outlet constitutes a temperature control zone. Self-powered VAV diffusers are most commonly used to create small sub-zones where it may be too expensive to add dedicated zones. For instance, VAV diffusers might be used in a small conference room served by an air conditioner also serving adjacent office spaces. They are typically available with heating/cooling change-over capability, controls that reverse control action depending on the supply air temperature, so the diffuser may be used on units that supply both heated

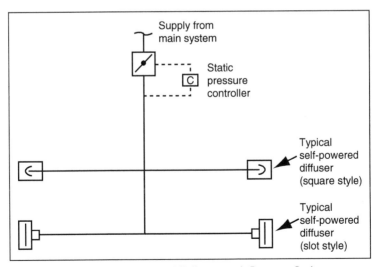

Figure 5-3   Self-Powered Diffusers with Pressure Reducer

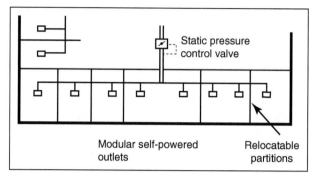

Figure 5-4   Modular Self-powered Outlets

and cooled air. Where many small zones are desired (such as in an application with many small private offices), VAV diffusers might be the best solution because they are generally less expensive than using small individual standard VAV boxes and conventional diffusers.

The self-powered boxes do not offer the control flexibility offered by other control methods. Perhaps the most obvious is that the occupant does not have the ability to change the set point. The set point can be changed but it requires access to the diffuser on the ceiling and often removal of the face plate. Compared to other control systems, it also lacks functionality as there is no ability for remote adjustment, night or unoccupied setback, or shut-off, or the ability for remote monitoring.

The limited pressure capacity and self-contained attributes can be attractive in specific situations. A simple example is a building which has many small spaces and a high frequency of changes in layout (high churn rate). A system such as shown in *Figure 5-4* can provide consistent temperature control for variable layouts. Changing layouts does not involve any change in wiring or tubing, and whatever the layout the space will contain its share of outlets each with internal control.

5.3 System-powered Controls

System-powered Air Valves

System-powered controls – devices that use supply air pressure as the power source – are a variation on self-powered controls used on early constant air volume (CAV) regulators and variable air volume boxes and diffusers. They were very popular in the late 1950s through 1970s, but they have largely been replaced by pneumatic, analog electronic, and digital controls in modern systems. While attractive because they do not require any external power source, system-powered controls often require that duct static pressure be relatively high (about 1–2 inch wg, depending on the manufacturer). This high pressure was standard on early high-pressure VAV systems, but not on modern, low-, and medium-pressure systems designed to minimize fan energy.

Some system-powered VAV boxes and VAV diffusers are designed for low pressure systems, requiring an inlet pressure of only about 0.5–0.75 inch wg, very close to that required for more conventional low-pressure VAV boxes. A typical application would be an air valve that, when there is a low static pressure, less force is applied to a cone inside a venturi, that makes the area larger, which allows more flow; and vice versa, during higher static pressure, the area gets smaller and again regulates the flow.

Each time, the opening changes (and therefore the pressure changes) to maintain its flow set point. While these devices do not incur the fan energy penalty common to most system-powered controls, they are usually limited to air-only applications as they cannot be used in conjunction with reheat coils or fan-terminals.

One advantage of system-powered controls over self-powered controls is that it is not necessary to maintain a low inlet static pressure. This is because the higher the inlet pressure, the higher the available power to drive the bellows or damper against that pressure.

Some system-powered controls are inherently pressure-independent. As pressure increases, the pressure in the bellows increases, reducing airflow correspondingly.

System-powered Water System Valves

There are many situations in water systems where the pressure or flow needs to be regulated at a fixed set point. Pressure can be regulated by, downstream of the valve, pressure reducing, or, upstream of the valve, pressure sustaining.

A pressure sustaining example is shown in *Figure 5-5*. There are three water-cooled units: A, B, and C. Water is pumped from the open tank through the units and drains back into the open tank. When the pump stops the system could drain. The pressure sustaining valve prevents that from happening. Note that in this example the valve works in the direction of flow. An on/off version of this valve is the safety, or pressure relief, valve. These valves do not allow any flow under normal circumstances as their relief pressure is well above normal operating pressure. Should overpressure occur, they open to let the pressure out.

If we wanted a valve to stop water draining back through the pump a simple check valve would be used. The check valve is designed with a plug that is

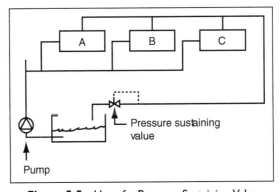

Figure 5-5   Use of a Pressure Sustaining Valve

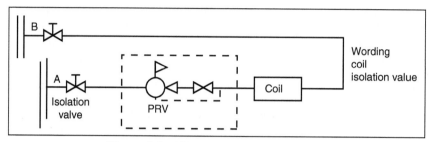

Figure 5-6   PRV and Control Valve Package

free to open easily with flow in the forward direction but to close completely if the flow reverses.

Maintaining the pressure downstream at a lower pressure is the task of a pressure reducing valve. Some equipment, commercial dishwashers for example, work best with a constant water pressure much lower than the typical city water supply. They would have a pressure reducing valve in their supply pipe to maintain the constant, lower, pressure.

Do you remember the section back in Chapter 3 on control valves and the challenge of varying pressure causing varying behavior? One way out of the problem is the have a pressure reducing valve in front of the control valve as *Figure 5-6*. The two valves can be purchased as a single, factory produced unit. The pressure reducing valve (PRV) maintains a steady pressure on the downstream side of the control valve. The pressure reducing valve effectively absorbs all changes in resistance due to the changes in flow (volume $\approx \sqrt{\Delta P}$) and control valve position. The result is a control valve that performs very close to its design characteristic.

The Next Step

In the next chapter, we will learn about electric control systems that are used primarily on simple systems, such as simple on/off control of fans and on small packaged HVAC equipment.

Chapter 6

Electric Controls

Contents of Chapter 6

Study Objectives of Chapter 6

Electric controls typically use 24 V ac, as a power source and use only contact closures (open-closed) and varying resistance (100–20,000 ohms), control logic; they do not use analog or digital electronics. The most common and basic temperature controller – the simple two-position thermostat – is an example of an electric control.

After studying this chapter, you should:

Understand how electric controllers (stats) are used to provide two-position control.
Understand how modulating controls work using bridge circuits.
Understand the difference between two-position, floating, and modulating actuators.
Know how to use electric controls in common HVAC applications.

6.1 Sensors

The temperature sensors most commonly used with two-position electric controls are the bimetallic strip, mercury switch, and the bulb-and-capillary, or remote bulb sensor, as shown in *Figure 6-1*. A bellows-style sensor is commonly used for modulating electric controls.

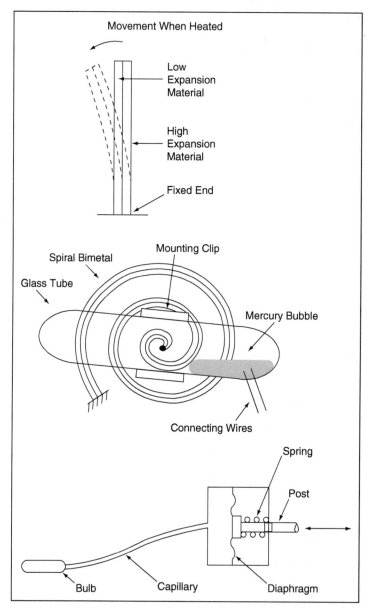

Figure 6-1   Bimetallic, Mercury, and Rod-Type Temperature Sensors

Humidity sensors used with electric controls are typically the mechanical type that uses a hygroscopic material (such as animal hair or a ribbon of nylon) that changes length as moisture is absorbed or desorbed into the material.

Electric differential pressure sensors primarily use Bourdon tube or diaphragm technology to switch a contact for two-position or floating control. Electric controls are seldom used for modulating control of pressure, because pressure is a fast-moving variable and usually requires a fast-acting signal. Similarly, flow status sensors are generally two-position switches using paddles or sails. (See Chapter 4 for a detailed description of sensors.)

6.2 Controllers, Two-position

Electric controls are most commonly two-position, using thermostats, humidistats, or pressure-stats wherein the controlled variable is sensed and compared to set point and a contact is opened and closed accordingly.

Figure 6-2 shows examples of how two-position and modulating controls are wired. In the first line, a cooling thermostat (one whose contact closes on a rise in temperature above set point) turns on and off a fan. If the fan was powered through the thermostat (if the round symbol in the figure was the fan motor itself), the thermostat would be called a line-voltage thermostat and its contact must be rated for the voltage and current required by the motor. A motor as large as about 1 horsepower (hp) can be controlled in this manner, as long as the contacts in the stat can handle the amperage and inrush current. Motors larger than about ½ hp typically have motor starters, with overload protection, and the round symbol in the figure would represent a starter coil or relay whose contact is wired to energize the starter coil.

The bottom line of *Figure 6-2* shows a floating controller that opens and closes a valve or damper through its actuator. When the controller senses that the controlled variable is above set point, the upper contact closes and opens the valve or damper. The actuator drives all the way open or closed, through its full stroke length, in a time period (usually 10–90 s). When the controlled variable is below set point, the lower contact closes, driving the valve or damper closed. When the controlled variable is within the controller differential, the switch is in the neutral position and the valve or damper stays in its last position. If you are not sure of how this works look back at Chapter 1, Section 1.4.

Packaged HVAC equipment has traditionally used electric controls, although many newer systems are supplemented with analog electronic or

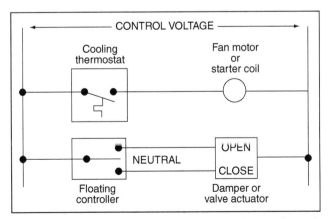

Figure 6-2 Typical Two-position and Floating Electric Controls

digital controls (see Chapter 10 for more details). In most cases, this equipment uses two-position or step control logic because the heating and cooling sources are inherently on/off (for example, staged gas valves on furnaces or staged compressors on direct expansion cooling units). When only two-position logic is required, controls are available that house a timeclock, and heating and cooling thermostats in the same enclosure. Some current packaged systems are now being outfitted with staged or modulating thermostats, with built-in humidity control, and modulating controls for Variable Frequency Dives (VFD for control of fan motor speed) and/or reheat controls.

An electric version of this control is shown in *Figure 6-3* applied to a typical packaged air conditioner with two steps of heating and cooling capability. Note the heat and cool anticipators (HA and CA) in the thermostat, which speed up the thermostat response (as you learned in Chapter 1). Note that the heating anticipation resistors are wired in series, so that they are on only when the corresponding stage of heat is engaged. The cooling anticipation resistors are wired in series so that they are on only when the stage of cooling

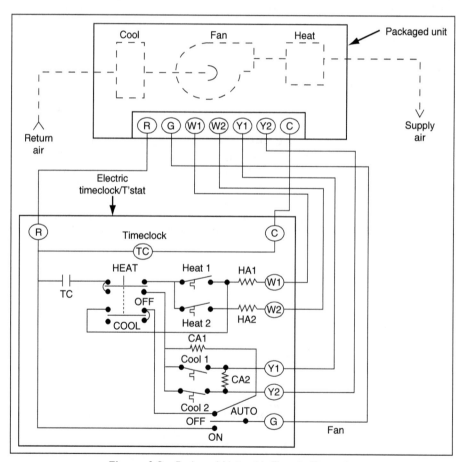

Figure 6-3 Packaged Unit with Electric Controls

is off, and you are in cooling mode. The cooling resistor effect is thus to make the thermostat think the temperature is rising more quickly that it really is and bring on the cooling earlier than if there were no anticipator heater. Note that the anticipation heater resistors turn off heating more quickly, but the cooling anticipator resistors turn on the cooling more quickly. Stage 2 cooling resistor is only on when it is in Stage 1 cooling and the second stage contact has not been made.

The controls in the package unit are not shown; these are factory mounted and wired by the manufacturer. Power is supplied to the packaged unit. A 24-V output transformer in the package unit provides power, via terminals R and C, to the electric timeclock/thermostat control circuit. Only the external devices and wiring need be shown with packaged unit controls. Manufacturers' drawings, provided with the unit, typically attached to the inside of access panels, show the details of the packaged controls.

Electric timeclock/thermostats are being used less and less. They have been replaced by programmable "electronic" timeclock/thermostats that are less expensive, more compact, are arguably easier to use, and most importantly have additional features such as timed override, ventilation control, cascaded humidity control, occupied and un-occupied set points, and set point setback/setup capability. These added features simplify the control design significantly by eliminating additional components and wiring. The wiring of a programmable thermostat looks exactly the same as an electric one, except that the devices in the controller itself are electronic rather than electric.

Step control is accomplished by typically using multi-stage controllers, which are basically a series of two-position controllers using the same sensor. Two and sometimes three stages of control are available. If more stages are required (for example to control a multi-stage refrigeration compressor), a stepping switch can be used (see *Figure 6-4*). It consists of a group of cam-operated switches mounted in a common enclosure. The camshaft is driven by a motor, controlled by a modulating controller. In effect, the stepping switch converts a modulating controller into a step controller. Angular adjustment of the cams allows the unit to close contacts one at a time until all are closed, or in sequence one at a time.

6.3 Controllers, Modulating

Electric controls are also used for true modulating control using the Wheatstone bridge, often referred to as a bridge circuit, shown in *Figure 6-5*. The bridge circuit consists of four resistors connected in a loop, as shown in the figure. A power source is connected across two diagonally opposite terminals. The other two terminals are connected to a load that may be a controlled device or may provide a modulating signal to an amplifying controller. The bridge circuit is used to send a varying output voltage to the load by varying the resistances.

The bridge circuit can be analyzed using Ohm's law ($V = IR$), which was introduced in Chapter 2. We know that the voltage drop through R2 and R3 must equal the voltage of the power source, typically 24 V or sometimes 120 V. Similarly, the voltage drop through R1 and R4 must equal this same voltage. We also know that the current through R2 is the same as that through R3 and the current through R1 is the same as that through R4. Expressing these knowns using Ohm's law (this is left as an extra exercise for the reader),

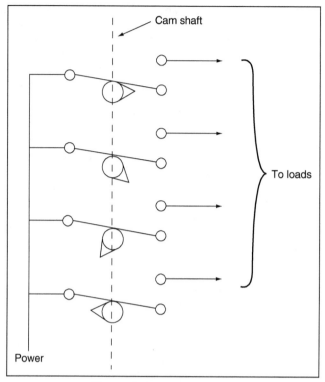

Figure 6-4   Stepping Switch

we can derive the following expression for the voltage across the output terminals:

$$V_0 = V_p \left(\frac{R2}{R2 + R3} - \frac{R1}{R1 + R4} \right)$$

(Equation 6-1)

where V_0 is the output voltage and V_p is the voltage of the power source. As this equation shows, varying any of the four resistors can vary the output voltage. If R2 is equal to R1, and R3 is equal to R4, the circuit would be balanced and, as can be seen from *Equation 6-1*, the output voltage would be zero. The output voltage is always less than the input voltage.

The bridge can be used in an electric controller as shown in *Figure 6-6*. The sensor is a variable resistor (such as a bellows modulating a potentiometer). The set point is adjusted with another potentiometer in the same part of the bridge circuit. To calibrate the controller when the sensor drifts, a third potentiometer is used, as shown in the figure. A more detailed example of how this bridge circuit is used with an actuator is discussed in the next section.

Three types of actuators are commonly used with electric controllers: proportional (modulating), floating, and two-position.

The proportionally controlled electric actuator is used in a bridge circuit (physical characteristics of which described in the previous section) to drive a

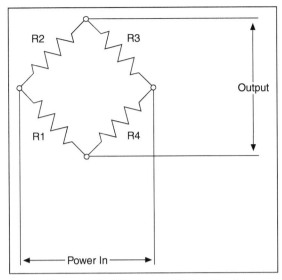

Figure 6-5   Wheatstone Bridge Circuit

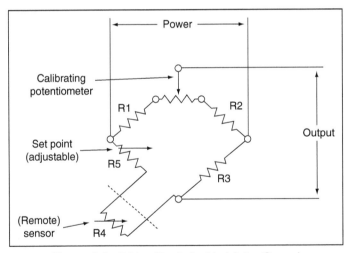

Figure 6-6   Bridge Circuit for Modulating Control

damper or valve using proportional control logic. One type of design is shown in *Figure 6-7*. The motor is reversible, driven one way or the other depending on the position of the yoke switch. A sensor-controller, shown on the right side of the figure, changes the position of a potentiometer in response to the offset from set point. The sensor potentiometer and the feedback potentiometer, which is driven by the motor, are divided by the wipers (depicted as arrows in the figure) into two segments each, thus forming the four parts of the Wheatstone bridge. The output of the bridge in this case is the current running from the controller to the feedback potentiometer. This current flows through a dual-coil

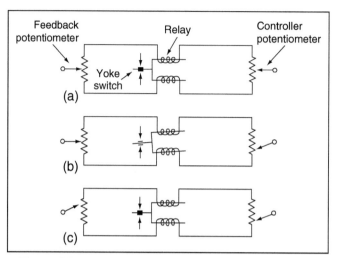

Figure 6-7   Modulating Three-wire Electric Control Motors

solenoid (shown as two spirals in *Figure 6-7*). The magnetic yoke arm mounted in the solenoid is tilted when the bridge is out of balance. The direction and extent of the tilt depends on the direction and extent of unbalance in the bridge.

In *Figure 6-7a*, the bridge is shown in the balanced position, with no current flowing and with the sensor wiper at mid-point and the valve or damper partly open. As the sensed variable changes, the sensor wiper will move, upsetting the balance of the bridge. Current flow in the control circuit will cause the electromagnets in the yoke to be energized, throwing the yoke switch to one extreme or the other and energizing the motor to drive the valve or damper accordingly (see *Figure 6-7b*). As the motor drives, the feedback potentiometer wiper will follow until the bridge is again in balance (see *Figure 6-7c*). At this point, control current ceases and the yoke switch returns to the neutral position, with the motor and the device it controls in a new position.

Had the bridge remained out of balance for a long period of time, the motor eventually would have driven all the way in one direction until it reached the limit of its stroke. At this point, a limit switch opens to keep the motor from locking and overloading, which would eventually cause the motor to fail. A clutch keeps the actuator from moving when the motor is disconnected in this manner. (Other means are now used to achieve this same motor protection, such as magnetic couplings and electronic current sensing circuitry.)

6.4 Example Application

Two-position electric controls are used in almost all control systems to turn on and off equipment. Chapter 2 goes through many examples of using on-off logic to accomplish various control sequences. This type of logic is used in conjunction with pneumatic, analog electronic, and digital control systems as well.

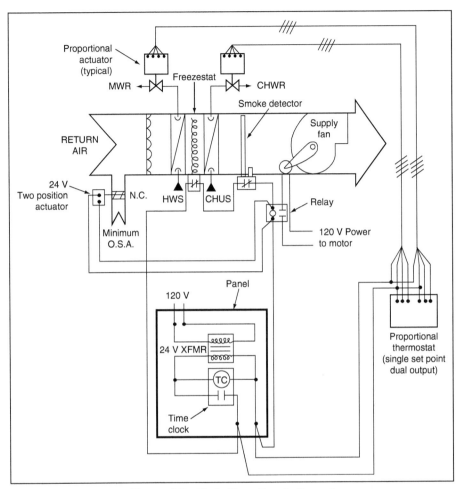

Figure 6-8   Single-Zone Electric Control

Figure 6-8 shows electric controls applied to a single-zone unit with hydronic heating and cooling coils. The control sequence for this system reads:

- Start/stop. The system shall start and stop based on the time schedule set on the seven-day timeclock. The fan shall stop if the freeze-stat indicates freezing temperatures or the duct smoke detector detects smoke.
- Heating/cooling control. Heating and cooling shall be controlled to maintain room temperature at set point (70–75°F, adjustable) by a single set point proportional room thermostat modulating two-way control valves.
- Outdoor air damper. The outdoor air damper shall open whenever the fan is energized and closed when the fan is off.

The system includes a small, temperature control panel housing a 24-V control transformer and timeclock. The panel could be located in a mechanical room for convenient but restricted access (to keep people from tampering with the timeclock, for instance). The fan is started using a relay rather than interlocking the fan motor's power wiring directly through the smoke detector, freeze-stat, and timeclock contacts. In this way, the contacts do not have to be rated for the current required by the motor. Also, the wiring from the panel to the remote devices is low voltage, which has fewer code restrictions than power wiring (120 V and above), and therefore should be less expensive to install. Note that power to the outdoor air damper is taken after the smoke detector and freeze-stat so that it will shut if either of these safeties trip.

The thermostat is a proportional thermostat with two potentiometers, allowing the two valves (or stages) to be controlled independently. The potentiometers can be adjusted so the valves operate in sequence.

Note that groups of parallel wires in *Figure 6-8* are lumped together and marked with hashes to indicate the number of wires, a common practice to make the drawing clearer and reduce clutter. This does not imply that the wires are electrically connected together, just that they are run next to each other to and from the same devices.

This basic design is shown. Some additionally desired control features are, for instance:

1. Night setback, which is often desired to keep the space from getting too cold when the unit shuts off at night. We would like the system to operate to keep the space warm, but we would also like the outdoor air damper to remain closed during this mode to reduce energy costs because the space is unoccupied.
2. It is often convenient to provide a bypass timer for occupants to override the time schedule temporarily. For this, a mechanical timer, typically 1–3 h maximum, can be connected across the timeclock contacts to turn the system on when it is wound up.
3. On systems that are served by a central chilled or hot water plant, that also serves many other systems operating on different time schedules, it is desirable to shut off the two-way valves when the unit is off to save pumping energy and reduce piping losses. This will require an additional power supply to the valve actuator which holds the return spring away from the valve shaft while power is on. When the power goes off, the spring is released, engages the shaft, and closes the valve.
4. Finally, as a freeze protection measure, the valves could be forced wide open in case the freeze-stat ever trips so that water will move through the coil, as moving water is less likely to freeze. This could be achieved by using a two-pole freeze-stat and utilizing the normally open contacts. If the freeze-stat operated, the contacts would close providing power to another relay which cuts power to the control thermostats and provides power to fully open the valves.

Each of these features is simple in concept but they significantly complicate the design. For this reason it is common practice for the designer to clearly identify the performance required from the control system but to let the controls vendor do the detailed circuit design and often make decisions about the valve/damper selection.

6.5 Actuators

Electric actuators may be a solenoid for two-position or motor driven for two-position or modulating control. Electrical actuators are available in two styles: direct-coupled actuators that directly connect without linkages to the controlled device (such as a valve or damper) and a general purpose, foot-mounted actuator that must be connected to the controlled device by linkages.

Actuators can be spring return or nonspring return. For valves that require the stem to be raised and lowered, typical for a globe valve, the actuator must provide a linear movement; hence, they are called linear actuators. The linear actuator must be able to apply an adequate force to operate the valve. The force is in pounds force (lbf). In most situations closing off against system pressure is the peak challenge and larger actuators will close off against higher differential pressures.

Rotary direct-coupled actuators are usually limited to medium-size valves and dampers requiring a medium amount of torque. They may either fit over the shaft and rotate or have a short arm stroke of between ½ and 1½ inches. For larger strokes, general-purpose actuators with linkages can stroke much greater distances. Direct-coupled actuators are usually less expensive and more reliable than the general-purpose actuator because of fewer moving parts and the linkage is eliminated. They are easy and quick to install, require no special tools, rarely require adjustments, and have low power needs.

General-purpose rotary actuators are useful when space requirements are limited, when greater torque is required, and larger strokes are needed. The torque ranges of general-purpose actuators are about the same as the direct coupled, about 10.00 inch lbf, but multiple actuators can be linked together on a shaft or through a linkage arrangement in order to overcome large torque forces. In addition, these actuators have been in the marketplace for a long time and they are therefore still in use for older retrofits.

Actuators are available with and without spring return. Spring return is used to provide a normal position of the valve or damper when power is removed (see Chapter 1). Actuators without spring return remain in their last position when power is removed. Many nonspring return actuators have a manual positioner device that will allow opening or closing of the actuator when power is lost. If no manual positioning device is available on the actuator, then manual valves or dampers should surround the controlled device so that the normal position can be open, and, if need be, manual devices can be closed off.

System gain is affected by the motor speed. Generally, these motors take 30–150 s (often adjustable in a range which may go up to 300 s or more) to travel from one extreme to the other. Motors of this type are made for use primarily with 24 V ac power sources, and, less commonly, with 110-V sources.

Floating-control actuators are similar to proportional actuators except they do not require the feedback potentiometer. (A feedback potentiometer can be added for monitoring purposes.) They are reversible and include limit switches as above.

With the typical electric actuator, the power source to the motor and the control power are the same. For this type of actuator, spring return cannot be used directly; the actuator cannot have a normal position. This is because if the actuator had a spring return, the damper would not remain fixed when the yoke or controller was in the neutral position. If spring return is required for a given control application, a special type of electric actuator must be used. These actuators have a separate power supply from the control signal. This second power supply powers a "clutch" that disengages the spring from the actuator when the power is on. When power is removed, the spring engages, returning the actuator to its normal position regardless of the control signal. This is one of the most widely used actuators in the marketplace today.

Two-position actuators generally are unidirectional with spring return, or nonspring return. Two-position motors are available for various voltages, most commonly 24/120 V ac but also for higher voltages such as 480 V ac commonly used to power large motors. These high voltage actuators are handy for opening dampers at fan inlets or discharges, with the actuator wired to the power circuit feeding the fan motor in parallel with the motor. In current applications, it is preferred to use transformers to step-down the voltage to 24 or 120 VAC in order to drive actuators. Normally actuators are slow moving, typically 60–150 s for their full travel, but for some applications it may be desirable to have them open and close faster, such as 5–15 s. One popular use for these actuators is opening and closing smoke dampers for life safety systems.

Auxiliary position switches are a common accessory to electric actuators. They close or open a contact when the actuator is in a certain position. Generally, the position is adjustable, although on some actuators the position is fixed to the end-positions, fully open or fully closed. One, two, and sometimes three auxiliary switches are typically available as options. As examples of its use, the auxiliary switch could be wired to start a fan only after a damper is nearly all the way open, or it could be used to turn off a pump when a valve is nearly fully closed.

6.6 Auxiliary Devices

Auxiliary devices used with electric controls were also discussed in Chapter 4. The main accessories for actuators are for chaining controls, or the provision of feedback. For example, in an earlier chapter we were discussing the use of a damper end switch to switch on the fan once the damper had opened. For this a single pole end switch would be required. If, instead we wanted feedback as to the damper position, we could specify a 5 k ohm (5,000 ohm), or 10 k potentiometer. This is a resistance coil with a slider which provides a resistance from 0 to 5 k ohms as the shaft rotates.

For situations where the damper actuator will be out in the weather, a weather shield may be specified.

Sometimes it will take multiple actuators to drive a damper; tandem-mounting kits can be used for this requirement. Also, as a damper ages, it sometimes requires more torque to operate because of wear, dirt, or other problems and may require more torque that can be supplied by multiple actuators.

While it is possible to control more complicated systems such as variable air volume systems using electric controls, more commonly these systems are controlled by pneumatic, analog electronic, or digital controls, and are the subjects of the following chapters.

The Next Step

In the next chapter, we will learn about pneumatic controls. Pneumatic controls were once the most commonly used controls in nonresidential applications, but they are largely being replaced by modern analog electronic and digital controls. Nevertheless, understanding pneumatic controls is important because they are still heavily used in existing buildings.

Bibliography

Honeywell Controls (2001) ACS Specialist CD-Rom. Minneapolis, MN.

Chapter 7

Pneumatic Controls

Contents of Chapter 7

Study Objectives of Chapter 7

Pneumatic controls, which use compressed air as the power source, are very simple and inherently analog, making them ideal for controlling temperature, humidity, and pressure. As such, they were once the most common controls used in nonresidential buildings, but they are rapidly being replaced by more capable and flexible analog electronic and digital controls (covered in the next two chapters). Pneumatic controls are still used commonly at the zone level (for example at VAV and CAV boxes, reheat coils, and fan-coils) and pneumatic actuators at valves and dampers are still commonly used in large digital control systems due to their reliability and cost. The movement of higher torque valves and dampers is especially cost effective using pneumatic actuators. There is also a very large installed base of pneumatic controls in existing buildings. For these reasons, it is important to understand how pneumatic controls work.

After studying this chapter, you should:

Understand the force–balance principle and how it is used in pneumatic controls.

Understand how temperature and pressure work in pneumatic control systems.

Understand how spring ranges can be used to sequence valves and dampers.

Understand how a controller's output and set point can be reset from other pneumatic signals.

Become familiar with various pneumatic relays such as averaging relays, selector relays, and reversing relays.

Know how to use pneumatic controls in common HVAC applications.

7.1 Principles of Operation

Pneumatic control systems use compressed clean, dry, and oil-free air as the source of control power, much like electric controls use electricity.

The compressed air, typically supplied in the range of 15–25 psig, is provided by an air compressor (discussed in Section 7.6). Because the air pressure can easily be varied, pneumatic controls are inherently modulating. The pressure is typically measured in pounds per square inch-gauge (psig) pressure units.

Basic in pneumatic control systems is the force–balance principle, shown in *Figure 7-1*. The enclosure or chamber is provided with three openings or ports, one each for the supply air input (also called main air), control signal output (to the controlled device), and exhaust. A flexible diaphragm allows an external sensor to vary the pressure on a lever arm. When the sensor presses down on the right-hand end of the lever, it lifts the left end upward, allowing the supply air valve to open. This increases the pressure in the chamber and also in the control signal output pipe, causing movement in a controlled device (not shown). As the pressure increases, it also *forces* the diaphragm up against the sensor pressure until the system is again in *balance* at a higher pressure with the supply air valve closed. Conversely, if the sensor action reduces the pressure on the diaphragm, then the spring on the left-hand end of the lever will force the right-hand end upward, allowing some air to exhaust out the exhaust port until the system is again in balance. *Figure 7-1* shows the principle. In practice, the actual details vary with the manufacturer.

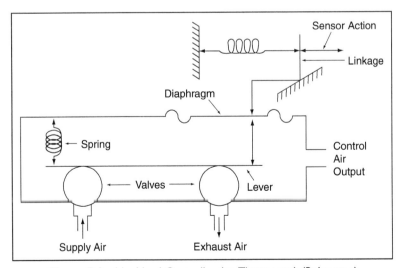

Figure 7-1   Nonbleed Controller (or Thermostat) (Relay-type)

The device shown in *Figure 7-1* can be used directly as a controller. The gain is adjusted by changing the length of the lever arm while the set point is adjusted by varying spring tension. Operation can be either direct-acting or reverse-acting (see Chapter 1), depending on the sensor action. The same force–balance principle is used to create pneumatic amplifiers, transmitters, and relays (which are discussed in Section 7.5).

Another common control device is shown in *Figure 7-2*. A sensor such as a bimetallic strip is used to open and close an air vent. As the bimetal changes position due to sensed temperature changes, it varies the vent opening area, thereby allowing more or less air to exhaust or bleed. A metering orifice called a restrictor is used to ensure that the rate of air that is exhausted can exceed the rate at which it is made up from the main air supply. If the air vent is left wide open, the air pressure to the output port (the signal to the controlled device) falls essentially to zero, although not completely to zero because of the continuous control air supply. When the vent is fully restricted, the output air pressure will build up until it is equal to the main air pressure. When the vent is partially restricted, a balance is obtained between the amount of air that is exhausted and supplied, causing the output pressure to be maintained at some value in between the two extremes.

The output (pressure signal) from the device in *Figure 7-2* is usually used as an input signal to a controller or indicator gauge. In this application, it is called a transmitter (in this case, a temperature transmitter). Pressure, flow, and other transmitters work on a similar principle. The device can also be used directly as a controller by adding adjustments for set point (spring tension) and gain (pivot point on the lever arm). It is a slow-acting controller because of the restricted air-flow rate to the device; this small amount of air is not enough to cause a rapid response at the controlled device.

The device shown in *Figure 7-1* is variously called a nonbleed, relay-type, high capacity, or two-pipe device, while that in *Figure 7-2* is called a bleed-type, low capacity, or one-pipe device. The term *bleed-type* refers to the fact that main air is delivered to the device continuously, and is constantly exhausted from it, except in the unusual condition where the bleed nozzle is fully shut. The term *low capacity* refers to the slow rate of change of the output signal due to the restricted main air connection. The term *one-pipe* comes from the fact that the device has only one connection (the control signal from the nozzle); the main air connection and restrictor shown in *Figure 7-1* are located elsewhere and are not a part of the device itself. On the other hand, the device

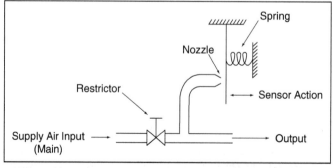

Figure 7-2   Bleed-type Controller (or Thermostat) (One-pipe)

shown in *Figure 7-1* requires two connections (main air supply and output signal), hence the term *two-pipe*; and it does not bleed air when the controlled system is steady, hence the term *nonbleed*.

Both one-pipe and two-pipe instruments can be used in any suitable combination in a given control system. The one-pipe device is simpler and less expensive to install, but it may not provide a satisfactory control response in some applications and in systems with long runs of air piping because of its inherently slow action. It also consumes more control air, increasing the operating cost, and often increasing the size of the air compressor. (Some relay-type controllers also bleed air at a continuous rate, depending on the manufacturer's design. However, the bleed rate is typically lower than a one-pipe controller.)

7.2 Sensors

Pneumatic sensors use the same basic sensing technologies described in Chapter 4 but are arranged to transmit a pneumatic signal generally in the range of 3–15 psig. In most cases, sensors are available with built-in relays to directly produce a pneumatic signal. Where this is not the case, transmitters or transducers (electric to pneumatic relays) can be provided that can convert an electric signal to a pneumatic signal. These are discussed in Section 7.5

In addition to the bimetal sensor shown in *Figure 7-2*, pneumatic systems make extensive use of the bellows sensors, bulb-and-capillary sensors, and rod-and-tube sensors discussed in Chapter 4. Rod-and-tube sensors are commonly used for sensing water in piping and for sensing duct temperature where there is uniform temperature across the duct. When duct temperatures are not uniform (for example in a mixed air plenum), a bulb-and-capillary sensor with a long bulb is preferred. The bulb is arranged in a grid pattern to read an average temperature across the duct.

Pneumatic humidity sensors in current practice are generally the dimensional change type using synthetic fabrics. They are arranged to send a signal to a controller through a nonbleed relay similar to temperature sensors. Flow and pressure sensors are the same as used with electric systems, but arranged to produce a pneumatic signal.

In pneumatic control terminology, the terms sensor and transmitter are used almost interchangeably to refer to a device that produces an output pressure that varies linearly to changes of a sensed variable. For example, a temperature transmitter designed to sense air or water temperature over a range from 80° to 240°F would have an output signal that varies proportionally from 3 psi (corresponding to 80°F) to 15 psi (240°F). The output signal range from 3 to 15 psi is the standard in the industry. To convert the signal from pressure (psi) to temperature (°F), the following equation (which applies to any transmitter with a 3–15 psi output signal) can be used:

$$T = \left(\frac{(T_H - T_L)}{(15 - 3)}\right)(P - 3) + T_L \qquad \text{(Equation 7-1)}$$

where T is the temperature corresponding to a pressure reading P, T_H is the high end of the range (240°F in this example), and T_L is the low end of the range (80°F in this example). The temperature from this transmitter corresponding to a 10-psig signal would then be:

$$T = \left(\frac{240 - 80}{12} \right)(10 - 3) + 80$$

$$= 173°F$$

To make this conversion easier, manufacturers typically provide conversion charts with the transmitter. If the temperature is displayed (such as on the face of a control panel), a pressure gauge would be connected to the signal fitted with a scale reading in "Deg. F" rather than psig. Pneumatic gauges often come with several scales designed to be used with standard transmitter ranges (for example, 35–135°F, 50–100°F, 0–100°F, 20–80% RH, etc.).

7.3 Controllers

Most pneumatic controllers (historically called receiver/controllers) are of the nonbleed type using force/balance principles described previously. The basic and by far the most commonly used controller is proportional-only. But, special controllers or relays are available that can produce proportional-plus-integral (PI) control logic. The output of pneumatic controls typically varies from 3 to 15 psig nominally, but the actual output can vary from zero psig up to the pressure of the main air supply.

The basic receiver/controller is commonly shown as in *Figure 7-3A* with three air connections (ports) for main air supply, sensor input connection from a pneumatic transmitter, and output to the controlled device. The labeling used to mark these ports varies by manufacturer. The main air connection is commonly labeled *M* for *main*, but it also can be labeled *S* for *supply*. The output signal is commonly labeled *B* for *branch*, but also can be labeled *C* for *control signal* or *O* for *output*. The sensor input is typically labeled *S* for *sensor* (conflicting with the manufacturers who label the main air connection *S*) or it is numbered, with the selection of numbered port varying depending on whether reverse or direct action is desired.

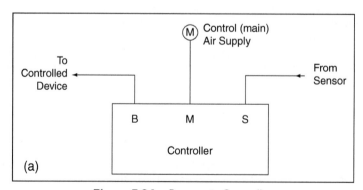

Figure 7-3A   Pneumatic Controller

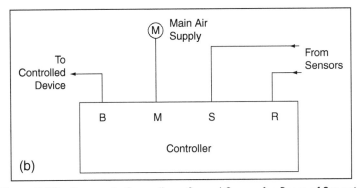

Figure 7-3B   Pneumatic Controller – Second Sensor for Reset of Set point

Numbered ports are also used with controllers with set point-reset capability (also called dual-input controllers), although the label *R* for *reset* is sometimes used. (See Chapter 5, Section 5.5 for an example of using a dual input controller for reset.) A dual input controller is shown in *Figure 7-3B*. Some controllers also have an input for a remote set point adjustment, thereby allowing set point to be adjusted from a remote control panel rather than at the controller itself. This port is labeled *A* for *adjustment*, *CPA* for *control point adjustment*, or simply numbered.

The set point can be manually adjusted with a dial on the controller. Remote set point adjustment, or adjustment from a remote sensor, is available for all dual input controllers. Control action may be direct or reverse, and is field adjustable. The proportional band setting is typically adjustable from 2.5 to 50% of the primary sensor span and is usually set for the minimum value that results in stable control. In a sensor with a span of 200°F, for example, the minimum setting of 2.5% results in a throttling range of 5°F (0.025 G 200 = 5°F). A change of 5°F is then required at the sensor to proportionally vary the controller branch line pressure from 3 to 13 psi. A maximum setting of 50% provides a throttling range of 100°F (0.50 G 200 = 100°F).

Reset authority, also called "reset ratio," is the ratio of the effect of the reset sensor compared to the primary sensor. The authority can be typically set from 10 to 300%. We introduced reset of boiler water flow temperature in Chapter 1. Suppose we wish to have the boiler flow termperature reset over the range 80–180°F (range 100°F) as the outside temperature drops from 60 to 10°F (range 50°F). For this situation 1°F change in outside temperature must reset the controller by 2°F, 200% reset ratio.

Room thermostats are the most common pneumatic controllers.

In addition to all-pneumatic control systems, they are used in conjunction with analog electronic and digital control systems, where pneumatic controls are used at the zone level (to control, for example, variable air volume boxes), while the more sophisticated electronic/digital controls are used at the system level (to control air handlers, chillers, etc.).

Many types of pneumatic thermostats are available, including:

- Single set point. This is the simplest of pneumatic thermostats available in both bleed-type (one-pipe), as shown in *Figure 7-2*, and nonbleed type (two-pipe), as shown in *Figure 7-1*. Both direct- and reverse-acting are available, shown schematically in *Figure 7-4* (see also Chapter 1). This thermostat is commonly applied to cooling-only or heating-only applications, but it

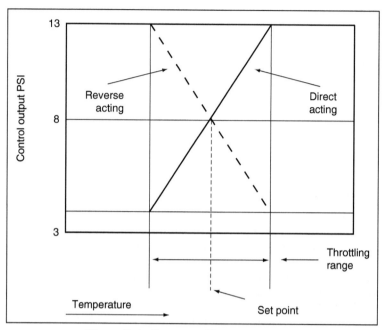

Figure 7-4   Pneumatic Thermostat Action

can also be used to control both heating and cooling in sequence by arranging the normal position and spring (control) ranges of the controlled devices. (Sample applications are discussed in Section 7.7.)

Deadband. This thermostat is used when the thermostat is controlling both heating and cooling and a deadband between heating and cooling set points is desired for energy conservation. Both bleed and nonbleed types are available. The output of a typical direct-acting deadband thermostat is shown in *Figure 7-5*. As the room temperature increases, so does the output up to about half of the output pressure range (typically 8 psig), enough to completely close the heating valve. As the room temperature continues to increase, there is no change in thermostat output over the (adjustable) deadband temperature range of perhaps 3–6°F. If the room temperature increases further, the thermostat output will again increase, opening the cooling valve.

- Dual set point. This thermostat provides the same deadband capability as a deadband thermostat by using two thermostats built into one housing (one for heating and one for cooling). Both bleed and nonbleed types are available. The deadband is created by adjusting the individual cooling and heating set points to create a gap in between (for example, setting the heating thermostat for 70°F and the cooling thermostat for 75°F). Each thermostat has a separate output that can be connected to separate controlled devices. Because the nonbleed version of this thermostat has three connections (main air supply and two outputs), it is often called a three-pipe thermostat (not to be confused with the day/night/vent thermostat discussed below, which is also sometimes called a three-pipe thermostat).

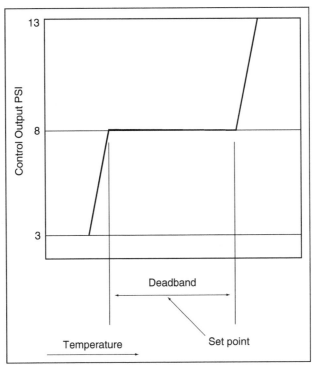

Figure 7-5   Deadband Thermostat Action

    The advantage of this thermostat over the deadband thermostat is that the control action of the heating and cooling devices can be mixed and matched as required by the application (for example, direct-acting heat/reverse-acting cool, reverse-acting heat/direct-acting cool, etc.). In contrast, the deadband thermostat can provide only one control action to both controlled devices. This dual control action capability provides a great deal of application flexibility without having to include additional devices such as reversing relays (discussed in Section 7.6). Another advantage is ease of recalibration because the dual set point thermostat is simply two standard thermostats, while adjusting and calibrating the deadband thermostat is more complicated and nonstandard.
- Dual pressure, reversing (heating/cooling). This thermostat is used when the controlled device changes from heating to cooling (such as a two-pipe change-over system), requiring that the control action of the thermostat change as well. The change in control action is initiated by raising the supply air pressure from 18 to 25 psig (actual pressures vary by manufacturer; 13–18 psig is also common). This would be done simultaneously with the change in controlled medium (such as the change from cold air to hot air or chilled water to hot water) supplied to the system controlled by the thermostat.
- Dual pressure, setback (day/night). Similar to the previous thermostat, this thermostat allows for a change in set point initiated by a change in control air pressure. Typically, when normal 18-psig control air is supplied, the thermostat controls to a comfort set point. When the pressure is raised to 25 psig

at night, the thermostat switches over internally to control to a set point that is lower (for heating systems), or higher (for cooling systems) to reduce energy usage when the space is unoccupied. (A more common means of achieving setback/setup on air systems is to simply shut the HVAC system off at night and use a separate night setback and/or setup thermostat to start and operate the system temporarily as required to prevent overly cold or hot space temperatures.) A variation on this thermostat is the day/night/vent thermostat, which has a second output in addition to the thermostat signal. This output is switched on (set equal to the main air pressure) when the thermostat is in the day mode, and switched off (the signal is bled off) at night. This auxiliary output was intended to open a ventilation outdoor air damper (hence the name, *day/night/vent*), but it can be used to initiate any day-only or night-only function.

Thermostats are available with blank locking covers; this is the most common arrangement because it discourages tampering by unauthorized users. Accessible set point adjustment and face mounted temperature indicators are also available.

Did you notice the words "blank locking covers; this is the most common arrangement because it discourages tampering by unauthorized users." We control temperature in the vast majority of situations for comfort. The ASHRAE Standard 55 on thermal comfort shows a temperature band of about 7°F for 80% of the occupants to be thermally comfortable assuming similar clothing.

So rather than allow adjustment to get as many people as possible comfortable we fix the temperature. Is it any wonder that people are not impressed with the engineering for comfort in many of our buildings?

Pneumatic thermostats and other pneumatic receiver/controllers require frequent recalibration to maintain accuracy. At least semi-annual (six-month) recalibration is recommended. Transmitters, such as temperature and pressure transmitters, should be recalibrated annually.

Despite this need for frequent recalibration, pneumatic controls are very durable and reliable. Almost all components (controllers, transmitters, relays, etc.) are interchangeable among manufactures. These advantages are often overlooked when control system type is selected.

7.4 Actuators

The pneumatic actuator (also called an operator or motor) is arguably the simplest and most reliable device used in HVAC control systems. As shown in *Figure 7-6*, it consists of a metal cylinder, a piston, a flexible diaphragm, and

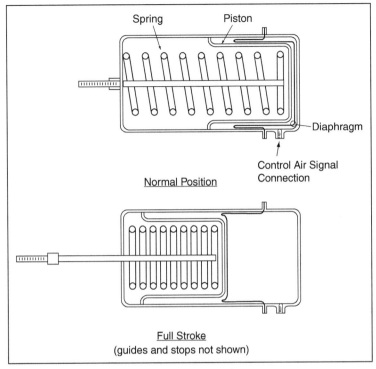

Spring   Piston

Diaphragm

Control Air Signal
Connection

Normal Position

Full Stroke
(guides and stops not shown)

Figure 7-6   Pneumatic Operator

a spring. In the normal position, with no air pressure, the spring is fully extended and the piston is at the top of the cylinder. When the control air pressure is increased, the diaphragm presses against the piston and spring, and moves the piston until the air pressure balances the spring tension.

The air pressure required to drive the operator through its full stroke is a function of the spring range. Many different spring ranges are available. A full range spring would typically have a 3–13 psig rating. More commonly, partial ranges are used; typically 3–8 psig, 5–10 psig, and 8–13 psig. These spring (control) ranges are used to sequence devices (as discussed in Chapter 1 and examples of which are discussed in Section 7.7 below).

The operator shown in *Figure 7-6* is a damper operator. A valve operator uses the same principles, but in a slightly different housing. The operator is connected to the valve or damper by a linkage. The way this linkage is connected determines whether the valve or damper is normally open or normally closed.

Hysteresis caused by friction, binding dampers, or linkages, aging, corrosion, and other factors, will prevent the valve or damper from responding to small changes in controller output pressure. Spring ranges can also drift, although this is not usually a problem. The pressure of the fluid being controlled (particularly in hydronic applications) will also cause the effective spring range to vary as this pressure increases to resist valve or damper opening or closure. These effects can cause control problems for modulating applications, particularly when spring ranges are counted on to sequence two or more devices.

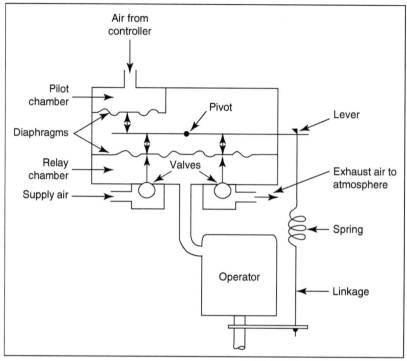

Figure 7-7   Positive (Pilot) Positioner

To mitigate their impact, a device called a positive (or pilot) positioner is recommended. This device, shown in *Figure 7-7*, is an amplifier and relay (using the force–balance principles discussed above) that causes the valve or damper to move to the position corresponding to the control signal regardless of the actual signal required by the actuator. A linkage connected to the shaft of the actuator senses its position. If the position is not that corresponding to the control signal, the positioner increases or decreases the actual signal to the actuator until it moves to the desired position. For instance, if a normally open valve is intended to be fully closed at 8 psig but the positioner senses that it is not, it will increase the actual signal to the actuator above 8 psig as required for the valve to fully close. Because the start and stop points of positioners are adjustable, they may also be used to offset the control range of an actuator from its built-in spring range to almost any control range, thereby allowing complex device sequencing or simply correcting for an actuator installation error.

7.5 Auxiliary Devices

In pneumatics, the term *relay* is applied to many functions, most of which bear no resemblance to the electromechanical devices described in Chapter 2. Relay functions include selection, discrimination, averaging, reversing, sequencing, amplifying, and switching.

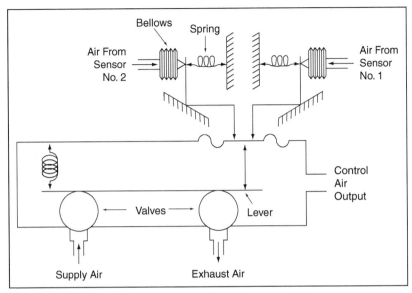

Figure 7-8   Higher of Two Pressures Relay

A signal selector relay is designed to select and pass on the higher (or lower) of two pressures entering the relay. The principle is shown in *Figure 7-8*, although current practice uses much simpler arrangements.

An averaging relay provides an output pressure that is the average of all input pressures. This is also used for set point reset, but it will not be satisfactory if loads vary greatly from zone to zone.

A discriminator relay, *Figure 7-9*, is used to select the highest (or lowest, or sometimes both highest and lowest) of several (typically six to eight) signals coming in to the relay. This device is often used in multizone or dual-duct systems to reset the hot and cold deck temperatures to satisfy the zone with the greatest demand, as shown in *Figure 7-9*. Its function can be duplicated by chaining together a string of selector relays, although the signal degrades somewhat with each relay so this approach is typically less accurate than using a discriminator.

A reversing relay reverses the incoming signal; a high incoming pressure will provide a related low outgoing pressure, and vice versa. For instance, if the incoming signal varies from 3 to 8 psig, the reversing relay output might vary from 8 psig (corresponding to the 3 psig input) to 3 psig (corresponding to the 8 psig input). The signal may also be offset and reversed. For instance, the 3–8 psig signal in the previous example could be converted to a 10–5 psig output signal, as shown in *Figure 7-10*. This is useful if one controller must control two valves or dampers, both of which are normally open but must act in opposite directions. (See Section 7.7 for example applications.)

For sequencing control, a ratio relay can be used. This multiplies or divides the signal in to provide a signal output that is a ratio of the input. For instance, a 3–8 psig signal may be broadened to a 3–13 psig signal to accommodate two control devices with 5 psig control spans.

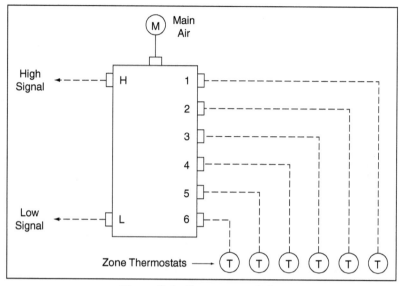

Figure 7-9   Discriminator Relay

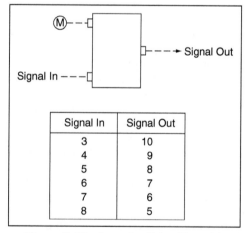

Figure 7-10   Reversing Relay (Example)

Most modern pneumatic control devices have very small ports and minimal air flow rates, but large volumes are sometimes needed for large operators. An amplifier relay provides a large volume at the same pressure as the low-volume signal coming to the relay.

A pneumatic switch (shown in *Figure 7-11*) is a two-position, three-way valve, switched by a set pressure signal to activate another control circuit. The pressure set point at which the valve switches positions is adjustable. A similar three-way valve, shown in *Figure 7-12,* operates on an electric signal rather than a pneumatic signal. Called an EP (electric-pneumatic) switch (or more correctly, an EP valve or

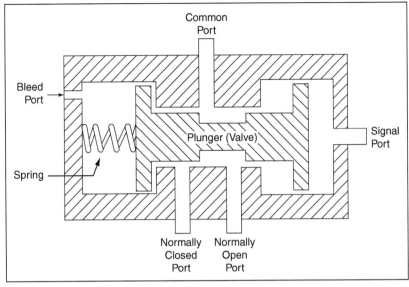

Figure 7-11   Pneumatic Switch

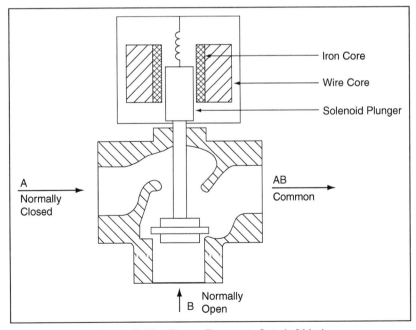

Figure 7-12   Electric/Pneumatic Switch (Valve)

solenoid air valve), it changes the position of the valve when the electromagnetic coil is energized by an electric circuit, typically 24 VAC/DC or 120 V.

It is important to note that pneumatic valves always are three-way valves even when the logical duty of the valve is two-position. This is because a pneumatic signal (pressure) will remain until it is bled off, unlike an electric circuit for which the electric signal is lost when a switch is opened. This concept can be better understood by the example shown in *Figure 7-13*. A damper is to be opened when a fan is turned on, as indicated by its starter auxiliary contact. When the contact closes, the common port of the EP switch is connected to the normally closed port, allowing main air to flow to the actuator, opening the damper. When the contact is opened when the fan shuts off, the EP switch is de-energized and the main air connection is shut off from the damper. But if only a two-way valve was used, the damper would remain open because the pressure in the line to the actuator would remain. This pressure must be bled off for the damper to close. Hence, a three-way valve is required with the normally open port (the port that is connected to the common port when the EP switch is de-energized) open to atmosphere. This is indicated as an arrow, as shown in *Figure 7-13*.

The connection of the three-way valve is important. Typically, the controlled device is connected to the common port. One port, either the normally open (NO) or normally closed (NC) depending on the wiring and piping arrangement, is typically connected to the output from a controller. The third port is usually either open to atmosphere or connected to main air to cause the valve or damper to fully open or close when the EP switch is energized (NC port) or de-energized (NO port).

Note that the terms *NC* and *NO* mean essentially the opposite of their meaning with electrical circuits. A port that is open to another allows control air to pass, while an open electrical contact does not allow current to pass. This can be confusing.

Another very common relay is the PE (pneumatic-electric) switch shown in *Figure 7-14*. It is simply a pressure actuated relay, usually with an adjustable set point, that will open or close an electrical contact. A typical application is to create step-control logic from a pneumatic controller (see Chapter 1). The PE is used commonly to interlock electrical equipment.

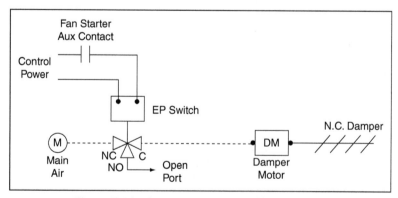

Figure 7-13   Electric/Pneumatic Switch Application

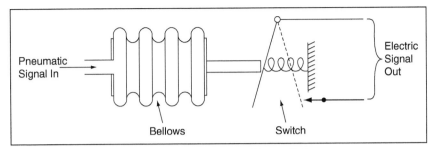

Figure 7-14   Pneumatic- Electric Relay

Another relay is used in air velocity measurement. The velocity pressure varies with the air velocity squared. As the velocity is doubled the pressure increases by $2^2 = 4$. A relay which outputs the square root of the input pressure is used to convert the input signal to a signal proportional to velocity. This device is commonly called a 'square root extractor'.

A transducer is a device for changing a modulating control signal from one energy form to another, keeping the two signals in proportion. The IP, or EP (current to pneumatic) transducer (shown in *Figure 7-15*) is used when an electronic controller is used with a pneumatic operator. The figure shows a solenoid in which the current may be varied, allowing the plunger to vary its distance from the nozzle of a bleed-type sensor. The output pneumatic signal will then vary in proportion to the input electronic signal. PI or PE (pneumatic to current) transducers are available to transmit a pneumatic sensor signal to an electronic controller. These transducers commonly use either a modulating current (1–20 ma) or voltage (1–10 VDC) signals.

Older pneumatic control systems made extensive use of manual switches. *Figure 7-16* shows a two-way switch, which is similar in effect to a single-pole/double-throw electrical switch (see Chapter 2). Many other configurations are available.

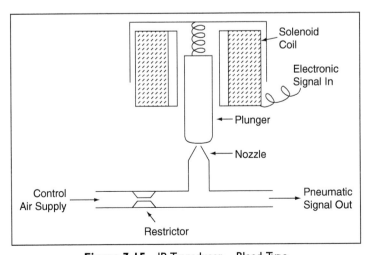

Figure 7-15   IP Transducer – Bleed Type

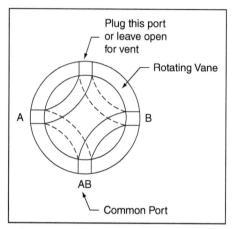

Figure 7-16   Manual Pneumatic Switch

Temperature, pressure, flow, and humidity can be remotely monitored from pneumatic sensors. The air pressure signal from the sensor is carried back to the desired monitoring point and used to drive a pressure gauge that has a display dial calibrated for the variable being monitored. Distance becomes a factor because air signals travel relatively slowly and can become attenuated over long distances. Maximums are about 1000 ft. Careful calibration and maintenance are required.

7.6 Compressed Air Supply

Pneumatic control systems require an adequate and reliable supply of clean, dry, and oil-free air. *Figure 7-17* shows the elements of such a system.

Most systems are supplied by duplex (dual) single speed, single stage compressors. Two compressors are desired for reliability because failure of the air

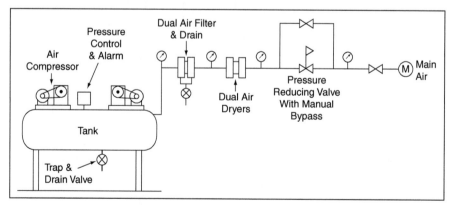

Figure 7-17   Compressed Air Supply

system will shut down the HVAC systems. Each compressor is sized to handle 1/3 to 3/4 of the calculated system control air requirement. The typical packaged compressor system has the compressors mounted on the air storage tank. For large systems, packaged units are not available and a separate tank must be provided. The tank is needed to provide storage to keep the compressors from short-cycling at low load. The compressors are controlled automatically to maintain storage pressure; usually with the two compressors alternated after each start to even wear and to allow the compressor to cool (some compressors have a tendency to leak oil into the air supply when their cylinders get hot).

Air is compressed to and stored at about 100 psi, then reduced to 20–25 psi for distribution. Pressure reduction and control is accomplished using a pressure reducing valve (PRV), which is a self-actuated control valve that maintains a constant leaving air pressure. It is recommended that many PRV stations be distributed around the building.

The air supply must be dry and oil-free. Therefore, air dryers and oil filters must be used, as well as particulate filters. Some oil or water may accumulate in the tank, requiring an automatic drain on the tank, controlled by a water level sensor. Oil-free carbon-ring compressors are sometimes used to minimize the possibility of oil contamination, but their maintenance costs may be higher and there is no general consensus as to their merit. A good oil filter will work satisfactorily provided it is regularly and carefully maintained. A coalescent-type oil filter offers improved performance by removing oil in both vapor and liquid form.

Air dryers may be refrigerated, removing moisture by cooling the air below its dewpoint, or desiccant dryers may be used, which remove moisture by adsorption. Both types require careful maintenance and oversight. Because failure of the air dryer can result in the destruction of all controllers, sensors, and most auxiliary devices, it is recommended to provide two dryers for redundancy with alarms to indicate failure.

Air requirements for normal operation are calculated from the manufacturer's data on consumption for the devices actually used in a given system. The calculation includes the number of devices, usage volume per device, and a use or diversity factor indicating the percentage of air volume expected at any one time. Bleed-type instruments use more air than nonbleed, so this is a consideration in selecting devices. The final sizing is usually done by the controls contractor, but the HVAC system designer must be able to estimate the horsepower requirements for the compressors for electrical coordination during design development.

Air piping has traditionally been copper. However, fire retardant plastic tubing is now available, with barbed or compression fittings, and is used extensively to reduce costs. Plastic tubing should not be used exposed, except for very short segments since there is a potential for damage, and can be fastened to other pipes/hangers, conduits or beams. It is most often used within enclosed control panels, enclosed within thin-wall conduit raceways, and in concealed yet accessible spaces (such as in ceiling attics and plenums). Where exposed to air systems (such as in fan or return plenums), fire-resistant (FR) tubing must be used to meet the flame- and smoke-spread ratings required by code. Copper piping may be run exposed and should be internally cleaned before installation.

Pipe sizing is based on the need to maintain a minimum pressure at controllers and controlled devices. For example, if the pressure leaving the PRV is around 25 psig and the minimum required at the most remote controller is

18 psig, then piping must be sized for a pressure drop less than 7 psi. This rather high range is typical and it leads to fairly small pipe sizes in most applications. Main air lines are typically 1/2 to 3/8 inch. Branch line tubing sizes of 1/4 inch and 1/8 (5/32) inch are most commonly used. Friction tables and charts are available from controls and air compressor manufacturers for sizing pipe for larger systems. To reduce pipe sizes in very large systems, distribution mains may be high pressure (50–100 psi) with pressure reducing valves (PRVs) as required to serve small areas of the facility (for example, one or two storeys of a high-rise building). Make sure the pneumatic main air lines and PRV locations are accessible and clearly identified.

As shown in *Figure 7-17* and other figures, main supply air connections are typically shown in diagrams as an *M* with a circle around it. This indicates a supply of air from the control air system. If the air is taken ahead of the pressure reducing assembly, it is often labeled as *HP* for high pressure. Frequently, control air is switched on and off for control purposes (such as to shut normally closed devices at night). In this case, the control air supply is labeled as *S* to indicate *switched* main air.

7.7 Example Applications

Figure 7-18 shows the symbols commonly used in pneumatic diagrams, although symbols can vary among designers and manufacturers. Because of this variation, each designer must provide a legend defining the symbols used in his or her diagrams. Unless the designer wants to use a particular control vendor, it is preferable to use generic symbols and indicate connections between devices by dashed lines. This allows the contractor maximum latitude in adapting a particular manufacturer's devices to the systems. Note that the following two figures, figures 7-19 and 7-20, do not conform entirely with the symbols here but the alternative symbols and explanatory notes make it clear what each item is on the drawing.

Figure 7-19 shows a single zone unit with hydronic heating and cooling coils, similar to the example introduced in Chapter 5 and developed in Chapter 6 with electric controls. As before, the control sequence for this system reads:

- Start/stop. The system shall start and stop based on the time schedule set on the seven-day timeclock. The fan shall stop if the freeze-stat indicates freezing temperatures or the duct smoke detector detects smoke.
- Heating/cooling control. Heating and cooling shall be controlled to maintain room temperature at set point (70–75°F, adjustable) by a single set point proportional room thermostat modulating two-way control valves.
- Outdoor air damper. The outdoor air damper shall open whenever the fan is energized and close when the fan is off.

The system includes a small temperature control panel to house an electric timeclock and EP valve. The panel could be located in a mechanical room for convenient but restricted access (to keep people from tampering with the timeclock, for instance).

The 120-V fan motor is started by running power wiring directly through the timeclock, smoke detector and freeze-stat contacts. These contacts must

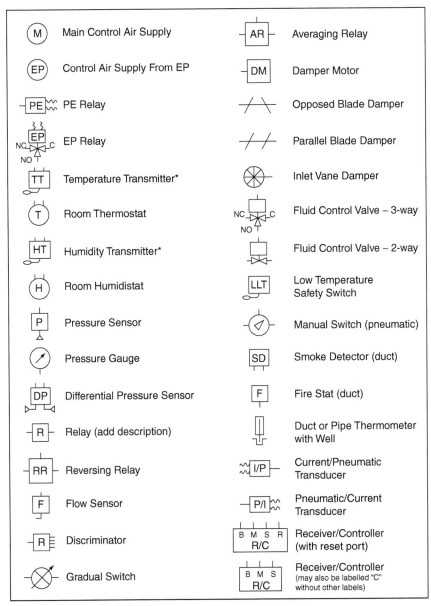

Figure 7-18   Symbols for Pneumatic Logic Diagrams

be rated for the motor current. This circuit also energizes the EP valve that provides main air to the outdoor air damper actuator, opening the normally closed damper. In this case, the EP coil must be rated for 120 V. Note that power to the EP valve is taken after the smoke detector and freeze-stat so that the damper will shut if either of these safeties trip. For slightly more foolproof performance, the EP valve could also have been powered in parallel with the

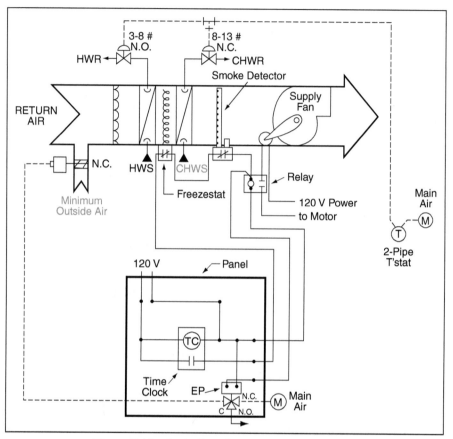

Figure 7-19   Single-Zone Pneumatic/Electric Control

fan motor from the load side of the starter. This would cause the EP valve to close the outdoor air damper if the starter drops out due to a motor overload (in addition to the other safety interlocks, such as smoke dampers).

The thermostat is a single set point thermostat controlling both the heating and cooling valves. For these valves to sequence properly, the control action and spring ranges must be properly selected. In this example, the heating valve was selected to be normally open so that even if the control air system failed, heat could be supplied to the space. This is a common fail-safe selection. The normally open heating valve requires that the thermostat be direct-acting so that a drop in space temperature will result in a drop in controller output signal, opening the hot water valve.

The standard spring range for a normally open valve is 3–8 psig. (Some manufacturers use 2–7 psig or 2–5 psig.) If the controller is to be used to directly control cooling as well, the cooling valve must be normally closed and operate from 8 to 13 psig. Again, this is the standard spring range available with normally closed valves.

As with the example discussed in Chapter 6, this design is missing some frequently desired control features such as night setback operating with the outdoor air damper closed, bypass timer for occupant override, shut-off of two-way valves when the unit is off, and forcing the valves open when the freeze-stat trips as a freeze-protection measure. These features are added in a manner very similar to the electric controls example, and it is left as an exercise to the reader to add these features to our pneumatic control system.

Figure 7-20 shows a variable air volume system with economizer. This system, introduced in Chapter 5, operates with the following sequence:

- Start/stop. The supply fan shall be controlled by the starter H-O-A (hand-off-auto) switch. When the H-O-A switch is in the auto position, the fan shall start and stop based on a seven-day timeclock. The fan shall stop if the freeze-stat indicates freezing temperatures or the duct smoke detector detects smoke, regardless of H-O-A position. Fan status shall be indicated by a current relay on fan motor wiring.
- Supply air temperature control. During normal operation, the supply air temperature shall be maintained at set point by sequencing the chilled water valve, economizer dampers, and hot water valve. The supply air temperature set point shall be reset based on outdoor air temperature as follows:

Outdoor air temperature	Supply air temperature
65°F	55°F
55°F	60°F

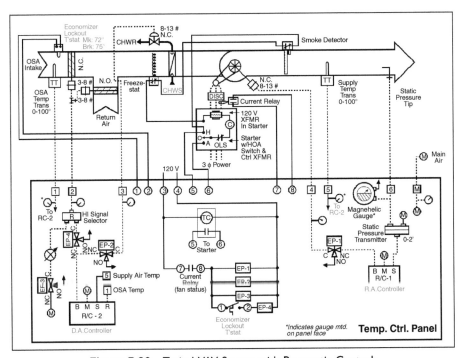

Figure 7-20 Typical VAV System with Pneumatic Controls

Both valves and the outdoor air damper shall be closed when the fan is off.

- Economizer. Economizer control shall be disabled when the outdoor air temperature is greater than the outdoor air high limit thermostat set point (67°F with a 3°F differential). The signal to the outdoor air damper shall be the larger of the signal from the supply air controller and the signal corresponding to minimum outdoor air intake required for ventilation (to be established by the balancing contractor in the field). (Note: this type of minimum ventilation control, while common, will not maintain a constant minimum amount of outdoor air and may not meet ventilation codes. More advanced ventilation outdoor air control designs are beyond the scope of this course.)
- Static pressure control. Whenever the fan is commanded on, duct static pressure shall be maintained at set point by modulating inlet guide vanes using PI logic. The static pressure set point shall be determined by the air balancer as that required to satisfy all VAV boxes when the fan system is operating at design air flow. Inlet vanes shall be closed when the fan is off.

The H-O-A switch is wired so that the normally closed contacts of the smoke detector and freeze-stat are wired before the hand (H) position so that the fan will not start if either safety contact is open. The auto (A) position is wired from the H contact through the timeclock contact to start the fan when the switch is in the auto position. The fan will then start either when the H-O-A switch is in the hand position, or when the switch is in the auto position and the timeclock contact is closed, provided the safeties are satisfied.

The normal positions of inlet guide vanes, chilled water valve, any duct smoke dampers, and outdoor air damper were all selected to be normally closed so that they will close automatically when the fan is off, as desired by the control sequence. This is done by bleeding the control signal from the line serving each of these devices using EP valves. When the fan stops, the current relay on one of the power lines to the fan motor will de-energize and its contact will open. As can be seen from the ladder diagram, this will de-energize EP valves EP-1, 2, 3, and 4, connecting their common port (which is piped to the controlled device) to the NO port, which is vented to atmosphere. This bleeds the signal, and the inlet vanes, chilled water valve, and outdoor air damper all close. Conversely, when the fan is on, the current relay contact closes which energizes the EP valves, thereby allowing the controller signal to modulate the controlled devices.

Note that control air supply is maintained to the two controllers in the panel. Two of the four EP valves in the panel could have been eliminated if we simply switched the main control air supplied to the controllers on and off with the supply fan. This would cause the control air to bleed out of the controllers when the fan shut off, and hence bleed the control signal lines to the normally closed valves, dampers, and inlet vanes. However, this was not done for two reasons:

- With many pneumatic controllers, it is an important design rule to maintain a constant control air supply because they tend to drift and require more frequent recalibration when supply air is intermittent. While some modern thermostats and controllers are generally resistant to this calibration drift, it is good practice nevertheless to maintain constant main air whenever possible.

- Bleeding air from the control valve and damper signals through the receiver/controller takes more time, and may not completely bleed the pressure depending on the specific design of the receiver controller. Any residual pressure may keep the valve or damper from fully closing.

While this design rule is recommended, it is commonly broken in practice without causing significant problems. A very common example is the use of a switched main air supply to zone thermostats. This might be done to effect a control sequence (such as to open up normally open VAV boxes for morning warm-up) or simply to save air compressor energy at night by eliminating the constant bleeding of air from the thermostats. If this is done, it is important that the thermostats specified be resistant to the calibration drift associated with intermittent main air supply.

The supply fan inlet guide vanes are controlled by sensing duct static pressure using a static pressure tip, which is simply a fitting that sticks out into the duct with a pickup designed to help ensure that static pressure in the duct is measured, rather than total or velocity pressure. This signal, which typically is in the range of 0–2 inch wg, must be amplified so that it can be used by a pneumatic controller. This is done using a static pressure transmitter.

Transmitters are available in various input signal ranges and it is important to pick a range that is as close as possible to the expected conditions so that the pneumatic signal is accurate and a large change in signal results from a normal change in static pressure. The transmitter output is fed into a reverse-acting receiver controller (RC-1) whose output is then piped to the normally closed inlet vanes. The controller is reverse-acting because upon a rise in duct pressure, we want the inlet vanes to close. Because the vanes are normally closed, they will close upon a drop in signal. This is in the opposite direction as the change in duct pressure, so the controller must be reverse-acting.

Supply air temperature is controlled by piping the signal from a supply air temperature transmitter to a direct-acting receiver controller (RC-2). The controller has a reset port that is connected to an outdoor air temperature transmitter. The controller is then adjusted to provide the desired reset schedule shown. This type of reset is common to reduce reheat losses in cold weather and to increase overall supply air rates for improved ventilation.

The output from RC-2 is piped to the chilled water valve and economizer dampers, which are arranged to operate in sequence. The outdoor air damper opens first when the controller signal is in the range 3–8 psig, then the chilled water valve opens when the signal rises to the range of 8–13 psig. Note that the spring range and normal position selections are important for proper operation.

The economizer damper signal passes first through an EP switch (EP-4) and a signal selector relay. The EP switch is energized whenever the fan is on and the outdoor air temperature is below 67°F, as indicated by an electric thermostat mounted in the outdoor air intake. The EP switch is piped so that the signal from the controller passes through the EP switch to the signal selector when the EP switch is energized. When the outdoor air temperature rises above 70°F (67°F plus the 3°F differential), the EP switch is de-energized, bleeding air from the signal to the signal selector relay. The signal selector is used to maintain a minimum position signal to the dampers whenever the fan is on. A gradual switch (analogous to a potentiometer in an electric control system) is adjusted as required to send the minimum position signal to the dampers as determined by the air balancer. The signal selector will then send the higher of the controller signal and the minimum position signal to the dampers.

Note that the pneumatic outdoor air temperature signal is not used to control the economizer enable/disable lockout. Theoretically, this signal could be sent to a pneumatic switching relay (see *Figure 7-11*). Instead, a separate electric outdoor air thermostat is used along with an EP valve. The reason is that the resolution of the outdoor air temperature signal is not likely to be fine enough to provide the 3°F differential desired (75°F less 72°F). For instance, the transmitter might have a range of 0–100°F, with 0°F corresponding to 3 psig and 100°F corresponding to 15 psig. This corresponds to 0.12 psig per °F (12 psig divided by 100°F). To achieve a 3°F differential, the switching relay must have a differential of 0.36 psig. Controllers with such a small differential are not generally available. Hence, the electric thermostat is used.

Pressure gauges are provided in key input and output lines in the panel to help in troubleshooting and as active displays of control points. Those gauges that are to be mounted on the panel face (such as outdoor air and supply air temperature and duct static pressure) are marked with an asterisk in the diagram.

As with the previous example, many optional controls could have been included to increase flexibility, but at higher cost and increased complication.

The Next Step

In the next chapter, we will learn about analog electronic controls. They are commonly used for systems that are small yet complicated; too complicated for electric controls and too small to justify the high overhead cost of pneumatic controls and some digital controls. Examples include packaged variable air volume unit controls. Like pneumatic controls, the falling cost of digital controls is reducing the usage of analog electronic controls. Nevertheless, understanding analog electronic controls is important because they are still used in existing buildings.

Bibliography

Honeywell Controls (2001) ACS Contractor CD-Rom. Minneapolis, MN.

Chapter 8

Analog Electronic Controls

Contents of Chapter 8

Study Objectives of Chapter 8

Modern electronic controls use analog semiconductor components to effect control logic. Several electronic controllers are commonly packaged into a single device that can handle all of the functions of a specific application such as control of a single zone air handler. Like pneumatic controls, electronic controls are being replaced by direct digital controls. In the pre-digital control era, electronic controls were most commonly used in packaged unitary equipment. However, they were never as popular as pneumatic controls for large commercial projects due to their higher cost, and lack of standardization. After studying this chapter, you should be able to:

Understand the function of operational amplifiers (op-amps) in electronic controls.
Become familiar with some of the application-specific electronic controllers available.
Know how to use electronic controls in common HVAC applications.

8.1 Principles of Operation

Modern analog electronic controls are based on semiconductor solid-state technology. They differ from electric controls, which are based on changes in resistance and voltage, using conventional resistors, bridge circuits, and potentiometers rather than solid-state devices. They also differ from direct digital controls that first convert analog signals into digital format before processing. Analog electronic controls use only analog signals and thus inherently can provide modulating control capability.

The term *analog electronic control* is often abbreviated in practice as simply *electronic control*, although, technically, digital controls are also electronic. The abbreviated term is used hereinafter for conciseness.

Electronic controls operate at variable voltages and currents, none that are totally standardized. Typical supply voltage is 24 vac or dc, requiring the use of a low voltage power supply composed of a control transformer, typically 120–24 vac, rectifier and conditioning electronics to provide 24 vdc. Input and output signals are either in the form of varying voltage or varying current. The most common current signal range is 4–20 ma dc (milli-amperes), although 0–7 ma dc and 10–50 ma dc are also used. When voltage provides the signal, it is usually 0–10 vdc or 0–16 vdc, although negative voltages are also rarely used. These ranges are analogous to the 3–13 psi output pressure signal range commonly used with pneumatic devices.

Electronic devices rely on microcircuits manufactured using deposition techniques similar to those used in computer chip manufacturing. These methods are continuing to improve, resulting in improved reliability and maintainability as well as lower costs. The Wheatstone bridge, which is discussed in Chapter 6, is also used in many of these devices.

Electronic devices can be interfaced with existing electric or pneumatic systems, using appropriate relays and transducers.

8.2 Sensors

All of the temperature sensors described in Chapter 4 can be used in electronic systems. The most commonly used temperature sensors are thermistors and RTDs. Some controllers are designed to accept a thermistor signal directly, with electronic circuits provided internally to convert the nonlinear resistance signal to a linear temperature signal. Others require an external transmitter that results in a standard 4–20 ma signal.

The most common humidity sensors are the resistance and capacitance types, packaged with electronic amplifiers to produce a 4–20 ma or other standard signal.

Virtually all of the pressure and flow sensors discussed in Chapter 4 can be used in electronic control systems. Again, all that is required is a transmitter to provide a standard electronic signal.

8.3 Controllers

Electronic controllers rely primarily on two fundamental devices: the Wheatstone bridge (described in Chapter 7), and the operational amplifier, or op-amp.

The op-amp is a solid-state amplifier that will provide a large gain while handling the varying input signals common to control systems. In the

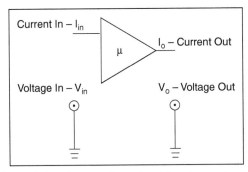

Figure 8-1   Idealized Operational Amplifier (Op-amp)

idealized op-amp (see *Figure 8-1*), the gain is the negative of the ratio of voltage-out to voltage-in:

$$Gain = -\left(\frac{V_o}{V_i}\right)$$ *(Equation 8-1)*

The inherent gain of an op-amp is very high. A small input voltage or current will produce a large negative output voltage. To make use of this device, the gain must be controlled.

Figure 8-2 shows a basic proportional op-amp arrangement. The feedback resistor allows most of the current to flow around the op-amp and back to the input. The negative output voltage from this feedback will cause the input voltage to drop to close to zero. The ratio of the feedback to the input resistors determines the gain.

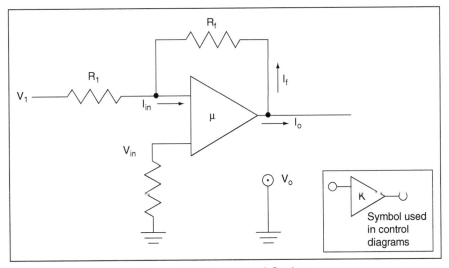

Figure 8-2   Proportional Op-Amp

For integral and derivative action (which are functions of time), a capacitor/resistor combination must be used, as shown in *Figures 8-3* and *8-4*. Time is required to charge the capacitor. To add or subtract, input summing arrangements are used.

Figure 8-5 shows a summing op-amp and *Figure 8-6* shows a subtraction op-amp.

An ideal controller using op-amps is shown in *Figure 8-7*. It begins with a comparison between the values of the set point and the measured variable, using a subtraction op-amp. This error signal is fed to the inputs of the three

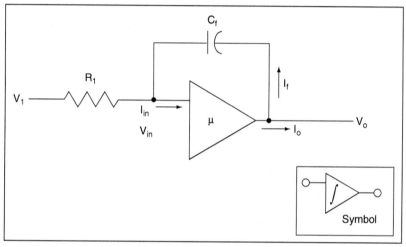

Figure 8-3   Integral Op-amp

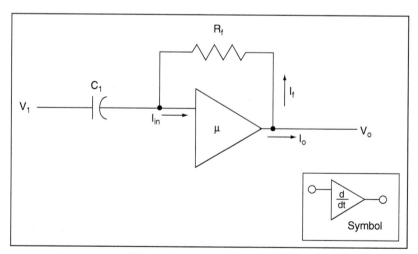

Figure 8-4   Derivative Op-amp

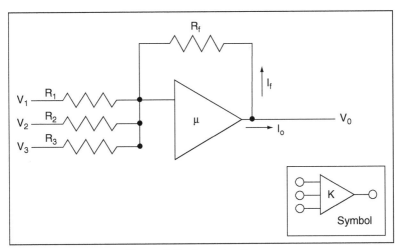

Figure 8-5   Summing Op-amp

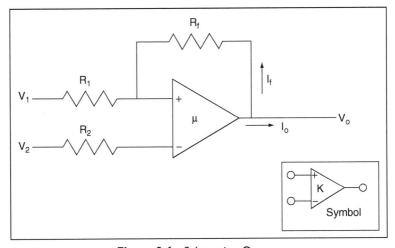

Figure 8-6   Subtraction Op-amp

mode op-amps, which provide proportional, integral, and derivative outputs. The three outputs are summed to provide the controller output.

All of these diagrams are highly simplified. Additional circuitry is needed to provide power, filter out unwanted noise, and provide stability. Adjustable potentiometers (adjustable resistances) are provided to adjust gains and set points. Switches may be provided to lock out the individual mode op-amps to allow any desired combination of modes.

While op-amps form the basic building block of electronic controls, the user seldom deals with these devices directly. Rather, op-amp based devices are packaged into individual controllers or multi-function controllers designed

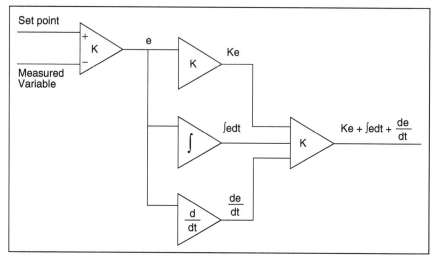

Figure 8-7   Idealized Op-amp Controller

for specific applications. Examples of application-specific electronic control-lers, often called logic panels, include:

- Single-zone unit controllers. These controllers include inputs for space temperature and, optionally, for supply air temperature, the latter to pro-vide master/submaster type logic (often incorrectly called anticipation by manufacturers of these controllers). Various output types are available, such as two or three stages of heating and cooling (for standard staged packaged unit compressors and heaters) and separate heating and cooling modulating outputs (for chilled and hot water applications). An econo-mizer output signal is also usually provided for use with externally supplied high limit lockouts and minimum position relays.
- Control of variable air volume air handlers and packaged units. These con-trollers have an input for supply air temperature and various output types similar to the single zone unit controller. The controllers typically are designed with a reset signal input for set point reset off outdoor air tempera-ture or zone sensors. An economizer output is also provided. These control-lers typically include adjustable set point, adjustable throttling range, and adjustable reset range. Integral control logic is usually standard to minimize proportional offset (droop). For staged outputs, time delays are built-in to prevent short-cycling: a very useful feature for cooling (compressor) systems.
- Control of multizone air handlers. These controllers provide hot deck, cold deck, and economizer control with optional reset from electronically con-trolled zone dampers. All set points, throttling ranges and reset ranges are adjustable.
- Economizer enthalpy control. These devices are used to shut off economi-zers when the outdoor air is no longer effective for space cooling. They directly determine enthalpy of air using relative humidity and tempera-ture measurements, without the need for other relays. They are available using differential logic (comparison of outdoor air and return air enthalpy) or fixed logic based on comparison of outdoor air conditions to set points that are curves when represented on psychrometric charts.

- VAV box controllers. Inputs include zone temperature and velocity pressure (differential between total and static pressure) for pressure independent control logic. Outputs typically include a floating-point output for the VAV damper. Heating outputs are either staged contacts for electric heat or modulating for hot water heat. Special controllers are also available for starting fans on parallel fan-powered mixing boxes.

Many other application-specific controllers are available. See Section 8.6 for an example of the use of these controllers.

8.4 Actuators

Actuators used for electronic control must accept one of the commonly used electronic signals such as 4–20 ma, 0–7 ma, or 2–10 Vdc. They can be divided into three categories:

- Electronic–electric. The actuator is an electric motor modulated by an electronic circuit receiving the control signal. Actuators are almost always powered by a 24-vac source. Some are available for 120 V service, but usually they operate on 24 V and simply include an internal transformer.
- Electronic–hydraulic. This actuator, shown in *Figure 8-8*, uses hydraulic pressure instead of an electric motor to achieve greater torque capability at low cost and low current.

  The small oil pump runs when the actuator is powered, typically by a 24 Vac power source. The oil pump creates a pressure available for the hydraulic work to be done. As the dc control signal varies from the controller, the transducer moves and restricts the flow of oil through the orifice, controlling the pressure on the diaphragm that drives the actuator shaft (piston) up and down.

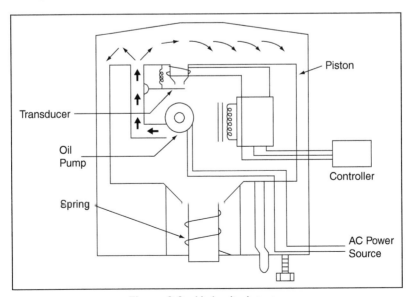

Figure 8-8   Hydraulic Actuator

- Electronic-pneumatic. In this hybrid application, pneumatic actuators are used, powered through an *I/P* transducer (Chapter 7). Many designers continue to use pneumatic damper operators with electronic controllers because of their simplicity, power, and reliability, but this is gradually changing as electronic-electric actuators become more common and less expensive.

Actuators are available with and without spring return. Spring return is a useful feature that allows the actuator to actuate to a desired position upon a loss of power since the spring will perform the work required to get to that position. For electric and hydraulic actuators, it is common for the actuator to include an integral dc power supply to power the control circuit.

One way electronic actuators differ from typical pneumatic and electric actuators is that the control signal and the actuator power supply are from separate sources. With most pneumatic and electric controls, the signal is also the source of power. The exceptions are pneumatic actuators with pilot positioners (discussed in Chapter 7), and electric actuators with spring return.

When using electric actuators, care should be taken into consideration about what voltage powers and forces the work of the actuator to be done, i.e. 24 vac or 120 vac. The issue to be careful about is the "VA" used by the actuator, and what size wiring you need to provide for it. For instance, for an actuator that needs 120 VA to operate, 1 amp to operate at 120 vac, and 5 amps to operate at 24 vac are required. Make sure the wire is sized appropriately for the amperage using the National Electric Code NFPA 70, or your local electrical code. And when using more than a few actuators on the same power source, be careful when using 24 vac, as there will be a voltage drop associated with the wiring. Problems may occur with control device proper operation if the supply voltage dips much below 22 vac.

8.5 Auxiliary Devices

Auxiliaries are similar to those used in electric systems. Relays will use solenoid coils designed for the low voltages used in electronics, 24 vac, with contacts rated as required for the higher currents and voltages required by motor starters and other electrical interfaces.

Relays are available that perform many of the same functions as pneumatic relays except using analogous electronic signals. For instance, relays are available for signal selection, signal reversing and signal sequencing. A relay similar to a PE switch (pressure electric switch) is available to convert an analog signal to one or more steps of control for two-position or step control logic.

Time clocks are electric motor driven devices that operate contacts when its time set points are met. Electronic versions of the electric time clocks use low voltage integrated timing circuits to operate its contacts.

8.6 Example Applications

A single-zone system with chilled water-cooling and a two-stage electric heater is shown in *Figure 8-9*. The supply fan, outdoor air damper, and associated controls are not shown for clarity. The control sequence for the system

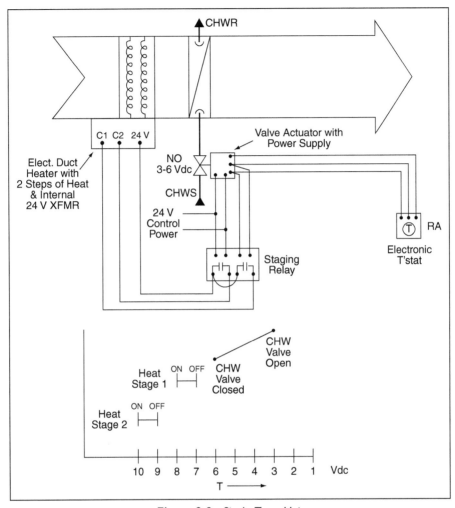

Figure 8-9   Single-Zone Unit

shown is simply to maintain the room temperature at set point by sequencing the electric heater stages and chilled water valve.

Control power at 24 vac is supplied to the valve actuator, which has an internal 24 vdc power supply to power the controls. This is a very common design with electronic controls. The dc power and signal are then wired to the thermostat. The signal is also wired to a relay that converts the electronic dc signal to discrete stages for staging the electric heater. The electric heater is supplied with its own control transformer (abbreviated XFMR), safety switches (such as an air-flow proving switch that keeps the heater off when there is no air flow), and contactors that energize the heating coil stages. Again, this is standard practice. The heater contactors are wired to the relay contacts of the staging relay. The heater's transformer can be used as the 24 vac source for the valve and staging relay if it is properly sized to have enough "VA" to power all of the devices.

For this design, it was desired that the heater should fail off if the control system power failed, keeping the electric heat elements from possibly over-heating. (In other cases, we might wish the system to fail with heat on for freeze protection. There are pluses and minuses to either approach.) If the heater must fail off, then the normally open contacts of the staging relay must be used, requiring that the thermostat be reverse-acting. This then requires that the cooling valve be normally open.

The positioner on the valve and the relay set points and differential are adjusted as shown in the *Figure 8-9* to provide the desired staging and sequencing. This is very much like pneumatic controls (with PE switches used for staging the heater) except the 3–13 psig signal is now a 2–10 vdc signal.

A single-zone logic panel "controller" (see Section 7.3) could also have been used to control this system. In place of the thermostat and staging relay would be the logic controller, which could be purchased with built-in modulating cooling output and multiple stages of heat, and a space sensor with set point adjustment potentiometer. An optional supply air temperature sensor could be used for master/submaster control. One advantage of using logic panels in place of individual electronic controls is that no adjustments are required to ensure that cooling and heating stages are properly sequenced, thereby simplifying both start-up and future maintenance of the system. In most cases, logic panels are also less expensive.

Figure 8-10 shows a variable air volume system with economizer operating with the same sequence as the example in Chapter 7, except using electronic controls.

An electronic timeclock is used instead of an electric timeclock (although either could be used in either application). This often requires the use of a relay to power the starter circuit (as shown in *Figure 8-10*) because the contact on the electronic timeclock is often rated for low voltage and low current.

An electronic logic panel, designed specifically for this application, is used to sequence the economizer and chilled water coil. If this were a packaged unit rather than a chilled water air handler, the controller would stage each step of cooling with built-in time delays to prevent short-cycling.

Power to the logic panel is shut off along with the supply fan. This is to keep the controller from PID integral winding-up (Chapter 1) when the sys-tem is off, resulting in poor control when the system is first started. This is a particularly important detail with direct expansion cooling systems, because the integral wind-up will usually cause the system to go to full cooling at start-up instead of staging up gradually.

The logic panel also controls the economizer, sequencing it ahead of the chilled water valve. A double-throw outdoor air thermostat enables the econ-omizer in cool weather and disables it in hot weather. A minimum damper position potentiometer is adjusted to maintain the minimum damper position required for ventilation regardless of economizer signal.

The actuators in this case are 24 V electric actuators with spring return. Like pneumatic actuators, they return to their normal position (outdoor air damper and control valve closed) when 24 V power is removed via the interlock to the current relay indicating fan status.

Static pressure is controlled in this example by an electric floating-point controller, which is less expensive than the equivalent electronic controller (another example of how many control systems are actually hybrids of the

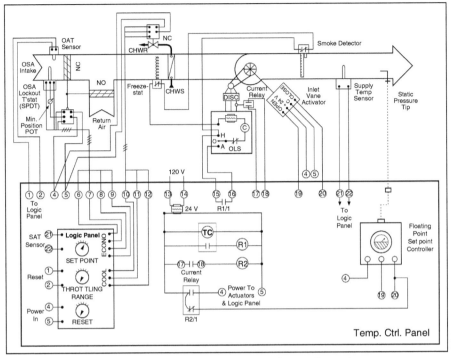

Figure 8-10 Variable Air Volume System Control

four primary control types). The controller drives the inlet vanes open or closed as required to maintain set point. The type of controller used has an integral pressure gauge for indicating both set point and duct static pressure. The controller is wired so that when the fan is off, the damper is driven closed. An electric actuator with spring return could also have been used, but this type of actuator is more expensive and requires additional wiring for the added power supply required to disengage the spring.

The Next Step

In the next chapter, we will consider control system design and how it can be written and drawn. Having shown the process steps examples will be used to illustrate their use.

Bibliography

1. Taylor, S. (2001) *Fundamentals of HVAC Control Systems*. Atlanta, GA: ASHRAE.

Chapter 9

Control Diagrams and Sequences

Contents of Chapter 9

Study Objectives of Chapter 9

This chapter introduces how control systems are designed and how the design intent is conveyed using control diagrams (schematics), damper and valve schedules, parts lists, and written control sequences. Electric controls are drawn on ladder diagrams, which were covered in Chapter 2, Section 2.7. The process of designing a control system must begin with an understanding of what is to be controlled and how it is to be controlled before selecting the type of control device hardware required to implement the desired logic. The hardware selection (be it self-powered, electric, analog electronic, pneumatic, or digital) will be a function of the requirements of the process, the accuracy and reliability required, cost desired, etc. More advanced control strategies are discussed in other courses and in the ASHRAE Handbooks.

After studying this chapter, you should be able to:

Comprehend design concepts that will result in a workable and flexible control system.
Understand the responsibilities of the various designers and contractors in the typical control system design and construction process.
Know how to create a control diagram using standard symbols.
Comprehend the guidelines for writing a control sequence.

9.1 Control Systems Design Criteria

Designing HVAC control systems, like designing HVAC systems themselves, is a science but allows for different approaches. Every designer's style differs and there are many right ways to control the same system.

The critical first step towards success is to be absolutely clear about the objective. For example, an office air-conditioning system has temperature, humidity, and outside air requirements to achieve during occupied hours. If groups of offices are to be in a single control zone identify the fact. It may well have requirements for setback temperature and humidity control in OFF hours as well as particular performance in case of fire/smoke alarm or fire department control. Having these requirements written down, and agreed with the client, makes for a solid foundation for controls design. This is particularly the case with large client organizations and larger design offices, where people with different views and expectations often join the project well after the design decisions have been made.

There are several basic requirements for systems:

- The control system must meet the needs of the process.
- The control system should control the process as directly as possible.
- The control system must be designed to work with the HVAC system, and vice versa.
- The HVAC control system should minimize energy consumption.
- The HVAC control system should maintain indoor air quality requirements.
- The cost of the HVAC control system must meet the budget.
- The control system must be designed for maximum simplicity. As Albert Einstein once said: "Everything should be as simple as possible, but not simpler."

The simplest systems have the best chance of being operated as intended. This is not to say that we must compromise the design goals expressed above for the sake of simplicity. The designer must balance the need for complexity with the reality that complex systems are the most likely to work incorrectly due to "bugs" in the design and the most likely to be misunderstood by the facility operator or repair technician who must make the system operate in the field. The ability of the maintenance team and available support should be considered when making design and equipment choices. A school which is one of many in a large city school district is very different from a school in a remote community, where maintenance expertise is very limited and any maintenance items may have to be flown in.

Even the simplest of control systems requires good documentation and training. Some control systems are necessarily complex because the HVAC system or the process it serves is complex. It is very important to have an effective training and useable documentation.

Operators that follow in years to come must be able to find out and understand how the system is designed to operate, and where components are physically located. Therefore, detailed explanations of the design intent should be included on plans. All software and programming must be turned over to the end-user and copies kept in the operations room as well as in another safe place. As-built drawings and data sheets must be transmitted through the design and contractor channels to the end-user files. Careful attention to depicting locations of all sensors, thermostats, communication lines, power wiring/tubing, pilot lights, valves, dampers, PRV's, relays, switches, etc., on the as-built and maintenance and operation documents should be done. All cables, wires, and tubes should be marked permanently with a marker, in at least two places, and documented on the control diagrams or point lists. It is a good idea for the engineer to specify and follow up to verify compliance. In any but the simplest system, formal commissioning pays off in initial performance and in ongoing operation and energy consumption.

9.2 Control Systems Design Process

The controls system designer must be able to convey not only the design intent (how the system is supposed to operate), but also who is responsible for performing the work and who is required to construct the control system.

Today, the advent of more complex control systems, and the variation in system architecture from one manufacturer to another, has led to more performance-based designs and specifications. In a performance-based design, the operational characteristics that are desired are specified and the control system provided can provide a number of solutions in order to control the system to meet this requirement. Ultimately, in most cases, the contractor is responsible for interpreting the intent of the contract documents, and the contractor shall direct his staff, subcontractors and suppliers to comply.

Controls Contractor: The controls contractor is part of the controls design-and-construct team, along with the HVAC engineer. The controls contractor may also be the control system vendor, the manufacturer, and/or distributor of the parts and components.

It is very important that the firm installing the electrical and mechanical controls is duly licensed and insured to do the work, as well as authorized and trained by the controls manufacturer. For additional information on the responsibilities and actions required at various stage of the design and construction process, readers are encouraged to read ASHRAE *Guideline 1, Commissioning HVAC Systems.*

9.3 Control Diagrams and Symbols

ASHRAE has published Standard 134-2005 Graphic Symbols for Heating, Ventilating, Air-Conditioning, and Refrigerating Systems, which includes standard symbols to be used in the controls industry: a sample of symbols can be seen in *Figure 9-1*. Many organizations have developed their own library of graphic symbols and it is wise to include on the controls drawings a key to their meaning to avoid any misunderstandings.

The symbols cover four sections: air handling, piping, controls, and equipment. The graphic symbol may have two types of information added about particular variations of device and detail about the device function. For example, a coil symbol may have CC – to indicate Cooling Coil – or HC – to indicate Heating Coil. Alternatively, the coil can be marked with a $\ominus$ for cooling

Controller	Symbol	Filename
Thermostat, wall mounted	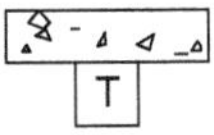	C 15970.001
Thermostat, duel mounted, low limit, manually reset	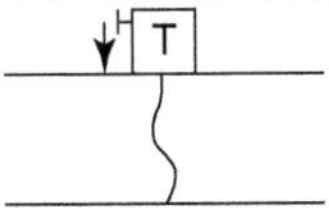	C 15970.005
Thermostat, duct mounted, averaging	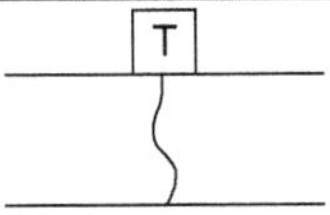	C 15970.002
Thermostat remote bulb	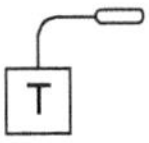	C 15972.003
Electric-preumatric transmitter	 ⟨ET⟩	C 15972.018
Electric-preumatric transducer	 EP	C 15960.001
Controller, adjustable (variable) speed drive	 ASD	C 15970.018

Figure 9-1   Sampling of Symbols > From ASHRAE Standard 134-2005

and ⊕ for heating. An example of indicating unction is with a thermostat, which may be marked as DA – Direct Acting – or RA – Reverse Acting.

Some of the graphic symbols are quite detailed and could be misread. Included in the figure is an example: thermostat, duct mounted, low limit, manually reset. Be sure to include the details in the specification or parts list so that you are not relying on detailed interpretation of the drawing to obtain what you want. Also note that not all symbols are unique, as in the abbreviation VFD for variable frequency drive and symbol ASD for adjustable speed drive, as shown in the figure.

When making up symbols, be sure to follow the overall convention of controller devices (thermostats, controllers) being in a square or rectangular box, and sensors and transmitters in a hexagonal box.

In this chapter, the diagrams are mostly generic; as they do not define the type of controls hardware (electric, electronic, pneumatic, and digital) and lines are used to indicate an interconnection between devices without defining whether they represent wiring or tubing.

Generic control diagrams are generally acceptable for most HVAC designs for bidding purposes because they, along with the written sequence of controls described in the next section, should sufficiently define the requirements of the control system. (The generic diagrams for digital control systems take on a slightly different appearance than the generic drawings here; if digital controls are to be used, diagrams should be tailored for that purpose and we will cover them in later chapters.)

9.4 Control Sequences

Controls are simply devices that try to duplicate the human thought process. With HVAC system controls, the controls are designed to carry out the thoughts and desires of the HVAC system designer. Because the HVAC system designer is probably not the person who will detail the controls system design, possibly not the person who will program and commission the system, and seldom the person who must operate and maintain the system for years to come, it is imperative that the designer convey his or her thoughts, about how the system is supposed to perform, clearly and without ambiguity. This is done through a written sequence of operation, also called a written control sequence.

Note that there are two fundamentally different ways of specifying a control system: performance and detail. In the performance specification the resulting performance is set out and described but the means to achieve this performance is left open. In the detailed specification, the components and how they are to be connected are all defined. "The temperature control shall maintain $73 \pm 2°F$" is a performance specification and "Supply and install a model ABC temperature sensor and XYZ controller" is a detail specification. Many designers use a mix of the approaches, allowing the contractor freedom except where the designer wants a specific product for performance, reliability, client wishes, or other reason. For the HVAC designer following the performance specification approach, a clearly written sequence of operation is essential because it is the means of conveying the desired control system's performance to the controls vendor or contractor.

Some guidelines for writing control sequences:

- Break the sequence into logical parts. First divide the system into major subsystems, such as air handlers, chiller plant and boiler plant, CAV, VAV, reheat coils, exhaust fans, isolation rooms, hoods, air flow monitors, etc. Then further subdivide each of these parts into individual control blocks, such as start/stop (how and when will the equipment run), and a block for each controlled device and/or controlled variable. In the previous chapter the variable volume system controls were in three parts: fan start/stop, air velocity control by inlet guide vanes, and temperature and mixed air control. Although the parts interconnected, they each operated as a separate subsystem.
- When specifying an HVAC control loop, be sure it is clear what the control point and controlled device are, and how the set point is determined. Typical devices to control are temperature, pressure, carbon dioxide level, humidity, carbon monoxide level, filter pressure drop, flow, and velocity.

If certain control logic (such as P, PI), resets, and direct/reverse acting are required for acceptable performance, it also should be specified.

- Many HVAC control loops will need to have some type of enabling and disabling interlock. This is important because PID control loops, which are enabled when the equipment they control are "off", can develop "integral windup" due to the long periods of time with its controlled variable signal is away from its set point. To solve this, the control loop can be turned on and off with a status control point tied to the status of the equipment controlled, allowing for the control loop to be turned on and off with respect to its equipment being on or off. This allows for the control loop to start working closer to its target set point immediately upon its equipment starting.

- Always state the required or desired set point and range for control loops. Many times, this information is left up to the operator to set because set points are generally adjustable. But the control sequence should at least provide a guideline of the range of set points that the system was designed to provide, and that will result in acceptable performance.

- In general, each controlled variable should be controlled by a single control loop. If various devices affect a given controlled variable and a control loop is provided for each, the control loops would have to be coordinated to be sure that only one is operational at any given time. For example, supply air temperature control for a system with an economizer should have a single controller that sequences the chilled water valve and economizer dampers. One controller should be used for each process. All sensors that directly control the controlled devices should be wired to that controller, where feasible. If two controllers are used instead of one, they must be enabled and disabled in sequence so that one does not "fight" the other. Some extra sensors may be required, that will wire to both controllers, if they directly conflict with its operation. Also, if one controller loses power or communication, or if sensors and/or devices wired to them affect other controllers, then this practice should be avoided. Generally, all control loops and sensors should have the capability to sense and alarm when there is a suspected and/or sensed failure and/or out-of-bounds response. In rare cases, where industrial high accuracy control is required, double loops or cascaded loops may be used.

- For each controlled device, state not only how it should be controlled when the system is normally operating, but also how the device is to be controlled when the system is shut off normally (such as by time schedule) or in an emergency (such as by a smoke detector, or fire alarm, or bad weather). For instance, when an air handler shuts off, typically it is better to have the outdoor air dampers to close to avoid infiltration and possible coil freezing. Also, two-way control valves to close to reduce pumping energy and piping losses may be desirable. If so, this must be specifically stated in sequences.

- Specify the normal position of a controlled device if it is important to the design. For instance, we may want all heating valves to be normally-open to prevent freeze-ups in cold weather. If so, this must be stated in sequences (or on schematics). If a normal position is not critical to the design, do not require one because with some hardware types (particularly electric and electronic controls) this will increase first-costs.

- Specify the wiring and installation methods required such as plenum wiring only (plenum wiring has low flame and smoke generation properties that are sufficiently limited for them to be approved for plenum use), conduit required, or a combination. Tell the contractor where to mount the devices, such as on the ducts, on the walls and at what height, on the top of the pipes, in the ceilings, etc.
- It is important that control sequences be as specific and detailed as possible. All sequences should answer the questions who, what, when, where, how, and how much?

9.5 Example Applications

In this section, written sequences and generic control diagrams are discussed for many typical HVAC applications. These applications, with some variations, will be the basis of further examples to demonstrate how different control system hardware and software types are applied.

These examples are written from the perspective of the HVAC system design engineer who is attempting to convey the design intent and operating sequence to a controls vendor/contractor (who will actually design more intricate details of the controls upon its formal submittal and as-builts) and to future system operators.

Typical Single-Zone System

Figure 9-2 shows a single-zone unit with hydronic heating and cooling coils. A fan coil unit can also be a simple single-zone system.

To develop a sequence of controls, the controlled devices, which are: the supply fans' motor; the hot water valve; the chilled water valve; and the outdoor air damper must be identified. The controlled variable is the temperature of the room being conditioned by this air handler. A room thermostat (a sensor and controller in the same enclosure) is used as the controller.

In addition, two safety devices are provided: a duct smoke detector and a freeze-stat. The smoke detector is required by most building codes (depending on equipment size) to shut off the fan when smoke is detected to prevent its recirculation to occupied spaces. The freeze-stat is typically required to shut down the fan and/or outside air damper in cold climates to prevent coil freeze-ups should the hot water system fail. For this system , the sequence might read:

- Start/stop. The system shall start and stop based on the time schedule set on the timeclock. (Modern timeclocks will have a total annual format with holiday and temporary day programming. Specify how many times each day that this start-stop occurs, such as on at 7 a.m. and then off at 8 p.m.) (If operation is desired after the normally scheduled "on" time, an "override or bypass" switch to allow is required for overriding and should be added to the diagram.)
- Heating/cooling control. Heating and cooling shall be controlled to maintain room temperature at set point (typically 70–75°F, adjustable) by a single set point proportional room thermostat modulating two-way

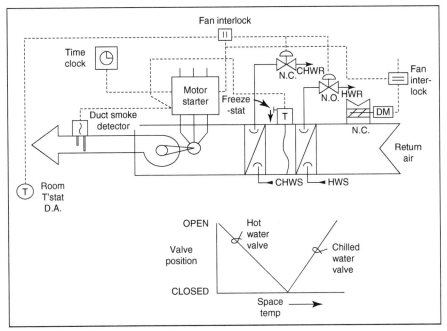

Figure 9-2   Typical Single-Zone Air Handler

control valves. On a call for cooling, the cooling coil control valve shall open. Upon a call for less cooling, the cooling coil control valve shall close. On a continued call for less cooling (call for heating) the heating coil control valve shall modulate open. When the fan is off, the hot water and chilled valves shall close.

- Outdoor air damper. The outdoor air damper shall open whenever the fan is energized and close when the fan is off. To maintain building pressurization, the exhaust fans for the general common areas and the toilets can be interlocked together with the operation of the OA damper. (Note: the exhaust fans should stay on and the OA dampers should stay opened during occupied periods.)

The control schematic does not show the wiring details of the fan starter, nor does it indicate the starter components required such as H-O-A (hand-off-automatic) switches and control transformers. They could be shown here, but it is usually more convenient to specify the requirements in a motor control diagram or in a generic starter detail. The timeclock should be mounted in a temperature control panel (TCP), and its physical panel location should be specified or shown on HVAC plans and in the specs. (If there are several panels, they are generally numbered (TCP-1, 2, etc.), as are control valves and control dampers (V-1, D-1, etc.).

The normal position of valves and dampers should be specified on schematics or in control sequences (but not both, to avoid potential conflicts). A schematic was selected here because it makes the written sequence more understandable. In this example, the hot water valve was selected to be normally open so that if the controller fails, for some reason, hot water flow could be made to flow through the coil by simply removing its control power. This might be used to prevent a coil freeze-up, or in cold climates it can be considered life safety. The controller must be direct-acting to correspond to the normally open hot water valve and the heating medium. This may or may not be included in the diagram and written sequence. If it were not specified, the controls vendor would be responsible for making the selection.

To correspond with the action of the controller above, the chilled water valve must be normally-closed. It also must have a control range that is sequenced with that of the heating valve. The control ranges may be shown on schematics, but their selection is often left up to the controls vendor or contractor.

For information, hot water pre-heat coils or electric heat is used to prevent the freezing of coils and other devices in cold weather. When heat is not available, sometimes just flow though the coils is initiated by the control system when freezing conditions apply, so that the coils can use the heat in the "piping system" to slow down the freezing effect. Heating reheat coils can also be useful to provide for dehumidification when needed and allowed by code.

On many systems, the designer will specify alarms that will advise the owner and operator that a control system or equipment failure has occurred, so a fast response and repair can be initiated.

In practice, many valves and actuators are now designed to permit the controlled device to be normally-open or normally-closed depending on what the designer wants.

The sequence specifies that the flow of hot and chilled water be shut off when the fan is not running, potentially reducing pump energy, piping, and coil heat losses. This could be done most conveniently by using normally-closed valves so that we need only interlock the valve's power source to the supply fan to close it when the fan is off. To do this, using a normally-open valve, some means of driving the valve closed when the fan is off must be provided. This would be a hardware requirement implied by specifying that the valve be normally-open but shut when the fan is off. (For cooling systems, it is advisable to shut off the cooling valve when the fan system is off, as it may produce condensate and waste energy when not being used.)

Using a single controller (thermostat) means we have a single set point for both heating and cooling. To save energy, we may wish to have a lower set point in the winter than in the summer. For instance, we may want to maintain the space at 70°F for heating and 75°F for cooling. Between these two set points, we would want both valves closed. This is called the controller deadband, which is the range over which a change in controlled variable causes no changes to any controlled device. To create a deadband using proportional controls, we could adjust the control ranges over which the chilled and hot water valves operated to create a gap in between them. Using pneumatic controls as an example, we could operate the hot water valve over a 2–6 psi range and the chilled water valve over a 10–14 psi range, leaving a gap from 6–10 psi where both valves were closed. The temperature range that corresponds to this signal range would depend on the throttling range of the controller. Typically, pneumatic thermostats are adjusted to provide a 2.5-psi per °F throttling range, which would mean the 4 psi gap between 6 psi and 10 psi would correspond to a temperature dead band of about 1.6°F, or 2°F, which could be 70–72°F.

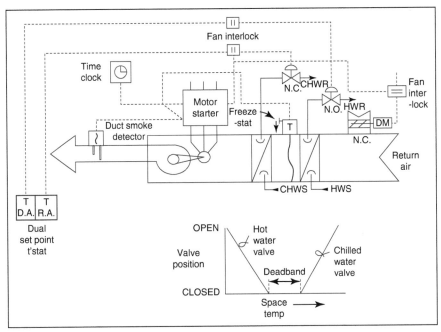

Figure 9-3   Single-Zone Unit with Dual Set point Thermostat

Another way to create a deadband is to use a deadband thermostat. This is a thermostat designed to provide a fixed output signal over a selected range of temperature that defines the deadband. Pneumatic deadband or hesitation thermostats are examples of this type of controller.

More commonly, simply using two thermostats, one for heating and one for cooling (as depicted in *Figure 9-3*), achieves deadband. Typically, the two thermostats are housed in the same enclosure. Common residential heating/cooling thermostats and dual set point electric or pneumatic thermostats are good examples of this type of controller. Dual output electronic controllers are similar except they generally use a single temperature-sensing element feeding both of the two controllers. Because each controller has its own set point, virtually any deadband can be achieved with this design. Unfortunately, in some cases it may be possible to overlap the two set points; for example setting the heating thermostat to 75°F while setting the cooling thermostats to 70°F. This would cause both valves to be opened at the same time, and the heating and cooling coils would "fight" with one another. To prevent this energy waste, many dual set point thermostats have physical stops that ensure that the cooling set point always exceeds the heating set point. (Digital controllers typically have similar limits programmed into software.)

Another advantage of using dual thermostats is that they can each have different control actions, which allow unlimited flexibility in selecting the normal position of the control valves. For instance, if we want both valves to fail open, we could use a direct-acting heating thermostat and a reverse-acting cooling thermostat.

Note that the problem of overlapping can also occur when two thermostats controlling separate equipment are mounted near each other. Two, or more,

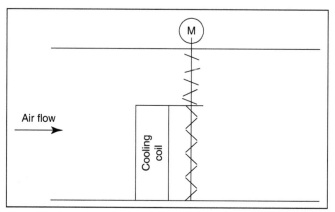

Figure 9-4 Face and Bypass Damper Arrangement

self-contained units that can both heat and cool installed in the same room is an example.

Typical Constant Air Volume System with Face and Bypass Dampers

For each air-handling unit, its chilled water valve shall be modulated to maintain discharge air temperature set point at 55°F (adjustable) unless it is in its heating mode.

Face and bypass dampers are used to vary the volume of air through a coil or bypassing it. The arrangement is shown in *Figure 9-4*. Due to the relatively higher resistance of the coil, the bypass damper is much smaller than the coil damper. The dampers here are shown as being driven by a single motor with the bypass well open and the coil flow somewhat closed off.

The face and bypass damper for this example is around the cooling coil. (Note, it can be around the heating coil or both for other designs.)

The face and bypass damper shall modulate to maintain a room or return air set point of 72°F (adjustable). As room or return air temperature rises above set point, the actuator shall command open face damper (more air flow across cooling coil) while closing bypass damper, until the desired return air temperature set point is achieved.

As the room or return air temperature falls below the set point 72°F (adjustable), the chilled water valve will be commanded closed, and face and bypass dampers shall be positioned to bypass position. On continued drop in room temperature, the face and bypass shall go to the full bypass position, and the hot water valve or electric heat shall be modulated or staged on as required to maintain heating set point.

Provide a minimum of 1°F of deadband where no control action occurs. Provide a fail-safe position for the face and bypass (F&B) damper to position itself. When the F&B damper is in the full face or full bypass position it is sometimes advisable to close the valve that is not being used. (Note, for example, in a cold climate, the fail-safe position may be in the bypass position.)

The supply fan and outdoor air damper/interlocked exhaust fans shall be controlled on and off together, on its own separate occupied time schedule.

Typical Constant Air Volume System with Multiple Zones or Reheat

The AHU is constant volume, and may or may not have inlet vanes or VFD to maintain its volume during filter loading. It typically has a main AHU cooling coil section that modulates in order to maintain its supply air temperature set point, often set at 55°F. In the reheat model, a thermostat in each room that is served reheats the cold air.

In the multi-zone model, a face and bypass damper arrangement, one for each zone, modulates its air stream through an electric or hydronic-heating coil and/or through the cold deck face, in response to a room thermostat/humidistat. Hot and cold air streams are mixed by the damper arrangement to control the zone temperature.

Typical Variable Air Volume System

The systems in the previous examples supply a constant volume of air to the space, with supply air temperature sometimes varying to satisfy space heating and cooling loads. Variable air volume (VAV) systems do the opposite. They maintain nearly constant supply temperatures while modulating the volume of air supplied to the zone.

Figure 9-5 shows a basic VAV AHU system which includes:

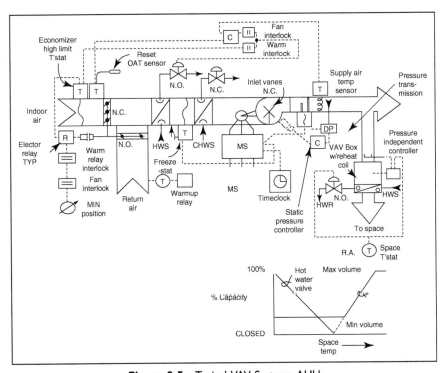

Figure 9-5 Typical VAV Systems AHU

- Supply air temperature control. During normal operation, the supply air temperature shall be maintained at set point by sequencing the chilled water valve, economizer dampers and hot water valve. The supply air temperature set point shall be reset based on outdoor air temperature as follows:

Outdoor air temperature	Supply air temperature
65°F	55°F
55°F	60°F

- Static pressure control. Whenever the fan is commanded on, duct static pressure shall be maintained at set point by modulating inlet guide vanes using PI logic. (VSDs are replacing inlet vanes in most cases these days.) The inlet vanes should be ramped very slowly in order not to cause undue turbulence or noise.

The pressure static high limit, PSHL, device stops the fan to avoid damage to the ductwork from a failed or closed damper downstream of the fan. Sometimes, if the vanes are not closed at shutdown, and the fan starts up under a full load this may happen prematurely. Also, if changes are made to the pressure relationships in the building, these high limit set points may have to be adjusted.

The PSHL is usually manual reset, and may initiate an alarm report. Take care to mount it, label it, and mark its location carefully, as it will need to be located someday for resetting.

The system in *Figure 9-5* uses inlet vanes for static pressure control, but the concept is the same regardless of which type of control device is used. Duct static pressure is measured in the main supply duct out near the extreme end of the system of VAV boxes. The further out in the system the sensor is located, the lower the set point can be, thereby reducing fan energy.

The static pressure set point is typically confirmed by the controls and air-balancing technicians with all VAV boxes open to their design maximum airflow rates. For systems with a large diversity factor (those for which the sum of VAV box design air flow significantly exceeds the fan design air flow) it may be necessary to reduce or shut-off the air flow to zones nearest the supply fan until the sum of the actual zone air flows approximates the fan design supply air flow rate. This will provide a more realistic approximation of actual operating conditions.

Some VAV systems use a return fan in addition to the traditional supply fan in order to effect building pressurization and deal with large pressure drops in return ducts. Controls need to work together in these supply/return fan systems in order to maintain their independent pressure set points and building pressurization goals.

PI control (proportional plus integral) is specified for static pressure control. In systems operating over a wide air volume range, the proportional gain must be fairly low (throttling range fairly high) to provide stable control. This can lead to offset (droop), as explained in Chapter 1, with proportional-only controls. The addition of integral logic can eliminate this offset.

- Economizer. Economizer control shall be disabled when the outdoor air temperature is greater than the outdoor air high limit thermostat set point (67°F with a 3°F differential). The signal to the outdoor air damper shall be the larger of the signal from the supply air controller and the signal corresponding to minimum outdoor air intake required for ventilation (to be established and verified by the balancing and/or commissioning contractor in the field).

The controller modulates the cooling control valve, economizer dampers and heating control valve in sequence to maintain the supply air temperature at a fixed set point. The economizer could be controlled by a separate mixed air controller, but using a single controller reduces first-costs and eliminates the complication of having to coordinate the actions of the two controllers so that they do not "fight" each other.

The sequencing of the components is accomplished by selecting sequential control ranges, shown schematically in *Figure 9-6*. The economizer function is very good about providing energy savings using outdoor conditions that are useful in helping to satisfy the cooling and heating loads inside the building. At the controller output signal corresponding to full cooling, the chilled water valve is full open. The economizer outdoor air damper may also be fully open depending on outdoor conditions. The control system will examine the outdoor temperature and humidity conditions, and use the appropriate amount of outside air to assist the conditioning of the space. (In the figure, it is assumed that the economizer high limit will shut off the economizer before full load is reached: the usual case.) As the cooling loads decrease, or when the set point is being satisfied, the chilled water valve begins to close. When it is fully closed after the load falls or the set point is completely satisfied, the signal to the economizer dampers, in sequence, falls, reducing the supply of cool outdoor air and therefore the space (or supply air temperature) becomes satisfied. As the load falls further, outdoor air is reduced until its minimum is reached after which the heating valve opens.

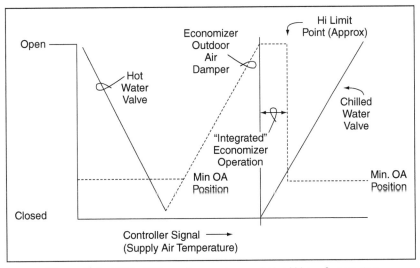

Figure 9-6 Chilled Water, Economizer, and Hot Water Sequencing

As explained in Chapter 1, sequencing outputs require both coordinating control ranges and coordinating normal position and control action. It is typically desirable for the outdoor air damper to be normally closed. In this way, the controller signal can be interlocked to the fan so that the damper will automatically shut when the fan is off. If the outdoor damper is normally closed, then the controller must be direct acting because, on a rise in supply air temperature, the output from the controller must also rise to bring in more outdoor air. The chilled water valve also must be normally closed because an increase in controller signal, which corresponds to an increase in the call for cooling, must cause the valve to open. The hot water valve then must be normally open.

For VAV systems such as this one, reheating of supply air will occur at the VAV boxes when zones require heat. To reduce this inefficiency, the supply air temperature set point may be reset. Various strategies have been used for reset control. The use of outdoor air temperature to reset supply air temperature is one of those strategies, used in this example.

This is shown graphically in *Figure 9-7*. Below 55°F outdoor air temperature, the supply temperature is constant at 60°F. Above 65°F, the set point is fixed at 55°F. In between these outdoor air temperature limits, the supply air temperature set point varies linearly from 55°F to 60°F.

Reset of outdoor air temperature is usually a reliable strategy, as long as interior zones, those that may still require cooling even in cold weather, have been designed for the warmer supply air temperatures that can occur due to the reset schedule (in this case 60°F). Care should be exercised in deciding on resetting strategies, as sometimes this resetting process can cause more problems than it solves.

The operations staff of the building and the control designer should be aware that relative humidity is affected by decisions about temperature (Harriman *et al* 2001). Consider hot and humid climates, where almost constant dehumidification is necessary almost all year long; when there is a reset, in these climates it is important to also look at the load in the building, the inside and outside humidity and the inside and outside temperatures, before there is a reset. Indoor air quality can be compromised if proper indoor temperature and humidity are not constantly maintained.

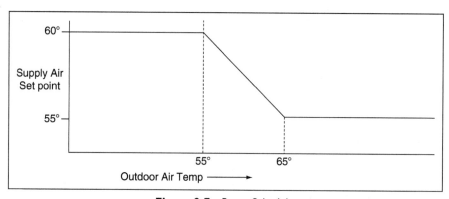

Figure 9-7   Reset Schedule

The above specification and discussion are written in narrative, descriptive language. Many organizations now use plain language – short simple sentences and bulleted items as the following example, which is modified from CtrlSpecBuilder.com. This system is similar to the variable volume system above with the addition of controlling the relief fan by building pressure compared to outside pressure and it has more alarms than we have previously considered.

Variable Air Volume – AHU – Sequence of Operations:

Run Conditions – Scheduled: The unit shall run based upon an operator adjustable schedule.

Freeze Protection: The unit shall shut down and generate an alarm upon receiving a freezestat status. Manual reset shall be required at operator panel.

High Static Pressure Shutdown: The unit shall shut down and generate an alarm upon receiving a high static pressure shutdown signal. Manual reset shall be required at operator panel.

Supply Air Smoke Detection: The unit shall shut down and generate an alarm upon receiving a supply air smoke detector status.

Supply Fan: The supply fan shall run anytime the unit is commanded to run, unless shutdown on safeties. To prevent short cycling, the supply fan shall have a user definable (1–10 min) minimum runtime.

Alarms shall be provided as follows:

- Supply Fan Failure: Commanded on, but the status is off.
- Supply Fan in Hand: Commanded off, but the status is on.

Supply Air Duct Static Pressure Control: The controller shall measure duct static pressure and modulate the supply fan VFD speed to maintain a duct static pressure set point. The speed shall not drop below 30% (adjustable).

- The initial duct static pressure set point shall be 1.5 in H_2O (adjustable).

Alarms shall be provided as follows:

- High Supply Air Static Pressure: If the supply air static pressure is above 2.0 in H_2O (adjustable).
- Low Supply Air Static Pressure: If the supply air static pressure is less than 1.0 in H_2O (adjustable).
- Supply Fan VFD Fault.

Return Fan: The return fan shall run whenever the supply fan runs.
Alarms shall be provided as follows:

- Return Fan Failure: Commanded on, but the status is off.
- Return Fan in Hand: Commanded off, but the status is on.
- Return Fan VFD Fault.

Building Static Pressure Control: The controller shall measure building static pressure and modulate the return fan VFD speed to maintain a building static pressure set point of 0.05 in H_2O (adjustable) above outside pressure. The return fan VFD speed shall not drop below 30% (adjustable).

Alarms shall be provided as follows:

- High Building Static Pressure: If the building air static pressure is 0.1 in H_2O (in H_2O is inches of water, just the same as 'in wg' that we used earlier in the text) (adjustable).
- Low Building Static Pressure: If the building air static pressure is -0.05 in H_2O (adjustable).

Supply Air Temperature Set point - Optimized:
- The initial supply air temperature set point shall be 55°F (adjustable) when the outside temperature is above 65°F.
- As outside temperature drops from 65°F to 55°F, the set point shall incrementally reset up to 60°F (adjustable) and remain at 60°F for temperatures below 55°F.

Cooling Coil Valve: The controller shall measure the supply air temperature and modulate the cooling coil valve to maintain its cooling set point.
The cooling shall be enabled whenever:

- Outside air temperature is greater than 60°F (adjustable).
- AND the economizer (if present) is disabled or fully open.
- AND the supply fan status is on.
- AND the heating (if present) is not active.

The cooling coil valve shall open to 50% (adjustable) whenever the freezestat is on.
Alarms shall be provided as follows:

- High Supply Air Temp: If the supply air temperature is above 58°F (adjustable).

Heating Coil Valve: The controller shall measure the supply air temperature and modulate the heating coil valve to maintain its heating set point.
The heating shall be enabled whenever:

- Outside air temperature is less than 65°F (adjustable).
- AND the supply fan status is on.
- AND the cooling is not active.

The heating coil valve shall open whenever:

- Supply air temperature drops from 40 °F to 35 °F (adjustable).
- OR the freezestat is on.

Alarms shall be provided as follows:

- Low Supply Air Temp: If the supply air temperature is below 50°F (adjustable).

Cooling Coil Pump: The recirculation pump shall run whenever:

- The cooling coil valve is enabled.
- OR the freezestat is on.

Alarms shall be provided as follows:

- Cooling Coil Pump Failure: Commanded on, but the status is off.
- Cooling Coil Pump in Hand: Commanded off, but the status is on.

Economizer: The controller shall measure the mixed air temperature and modulate the economizer dampers in sequence to maintain a set point 2°F (adjustable) less than the supply air temperature set point. The outside air dampers shall maintain a minimum adjustable position of 20% (adjustable) open whenever occupied.

The economizer shall be enabled whenever:

- Outside air temperature is less than 65°F (adjustable).
- AND the outside air enthalpy is less than 22 Btu/lb (adjustable)
- AND the outside air temperature is less than the return air temperature.
- AND the outside air enthalpy is less than the return air enthalpy.
- AND the supply fan status is on.

The economizer shall close whenever:

- Mixed air temperature drops from 40°F to 35°F (adjustable)
- OR the freezestat is on.
- OR on loss of supply fan status.

The outside and exhaust air dampers shall close and the return air damper shall open when the unit is off.

Minimum Outside Air Ventilation: When in the occupied mode, the controller shall measure the outside airflow and modulate the outside air dampers to maintain the proper minimum outside air ventilation, overriding normal damper control. On dropping outside airflow, the controller shall modulate the outside air dampers open to maintain the outside airflow set point (adjustable).

Humidifier Control: The controller shall measure the return air humidity and modulate the humidifier to maintain a set point of 40% rh (adjustable). The humidifier shall be enabled whenever the supply fan status is on.

The humidifier shall turn off whenever:

- Supply air humidity rises above 90% rh.
- OR on loss of supply fan status.

Alarms shall be provided as follows:

- High Supply Air Humidity: If the supply air humidity is greater than 90% rh (adjustable).
- Low Supply Air Humidity: If the supply air humidity is less than 40% rh (adjustable).

Prefilter Differential Pressure Monitor: The controller shall monitor the differential pressure across the prefilter.

Alarms shall be provided as follows:

- Prefilter Change Required: Prefilter differential pressure exceeds a user definable limit (adjustable).

Final Filter Differential Pressure Monitor: The controller shall monitor the differential pressure across the final filter.
Alarms shall be provided as follows:

- Final Filter Change Required: Final filter differential pressure exceeds a user definable limit (adjustable).

Mixed Air Temperature: The controller shall monitor the mixed air temperature and use as required for economizer control (if present) or preheating control (if present).
Alarms shall be provided as follows:

- High Mixed Air Temp: If the mixed air temperature is greater than 90°F (adjustable).
- Low Mixed Air Temp: If the mixed air temperature is less than 45°F (adjustable).

Return Air Humidity: The controller shall monitor the return air humidity and use as required for economizer control (if present) or humidity control (if present).
Alarms shall be provided as follows:

- High Return Air Humidity: If the return air humidity is greater than 60% (adjustable).
- Low Return Air Humidity: If the return air humidity is less than 30% (adjustable).

Return Air Temperature: The controller shall monitor the return air temperature and use as required for set point control or economizer control (if present).
Alarms shall be provided as follows:

- High Return Air Temp: If the return air temperature is greater than 90°F (adjustable).
- Low Return Air Temp: If the return air temperature is less than 45°F (adjustable).

Supply Air Temperature: The controller shall monitor the supply air temperature.
Alarms shall be provided as follows:

- High Supply Air Temp: If the supply air temperature is greater than 60°F (adjustable).
- Low Supply Air Temp: If the supply air temperature is less than 45°F (adjustable).

Of particular note in this plain language specification is the use of AND and OR logic, Boolean logic:
The economizer shall be enabled whenever:

- Outside air temperature is less than 65°F (adjustable).
- AND the outside air enthalpy is less than 22 Btu/lb (adjustable)
- AND the outside air temperature is less than the return air temperature.
- AND the outside air enthalpy is less than the return air enthalpy.
- AND the supply fan status is on.

The economizer shall close whenever:

- Mixed air temperature drops from 40°F to 35°F (adjustable)
- OR the freezestat is on.
- OR on loss of supply fan status.

This style of presentation is very clear and easy to follow, particularly where many control choices are being made.

CtrlSpecBuilder.com is a website providing noncopyright specifications and drawings for common equipment based on direct digital controls. You can use it as a base for either DDC or non-DDC controls. Let us now return to the descriptive style and move from the VAV supply system to the VAV box. *Figure 9-8* shows a control sequence for a VAV box, with a detailed discussion of the philosophy behind each item.

- VAV boxes. Room temperature shall be controlled by modulating the air volume damper and the reheat valve, or electric heater, in sequence as indicated in the *Figure 9-8*. Volume shall be controlled using a pressure-independent controller. The maximum and minimum cooling volume, as well as a heating minimum, shall be as indicated on equipment schedules. For units with electric heat, care should be taken to specify an air-proof switch (so the heater cannot come on without airflow to avoid overheating),

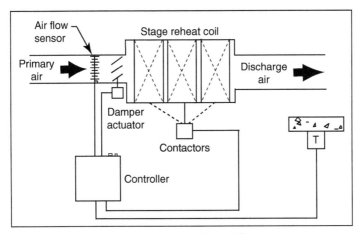

Figure 9-8 VAV Box Control Diagram

as well as thermal protection at the electric heater, so that potential fire conditions can be avoided. Power for the control system may be wired via a low voltage transformer attached to the electric heater, or, in the case of hot water heat, the power would be provided externally. This sequence references pressure-independent control.

Early "pressure-dependant" VAV systems simply used a damper mounted in the ductwork that was directly controlled by the space thermostat. As space temperature increased, the VAV damper opened. As the temperature in the space fell, the VAV damper closed until at zero loads (when the thermostat was satisfied) the VAV box would shut completely, with some (and sometimes, significant) leakage. If a minimum flow was required, the damper could be linked so that it would not close all the way to shut-off.

This type of control is called pressure-dependent because the amount of air supplied to the space is a function of the pressure in the supply air system, not just of the thermostat signal. A change in system pressure could cause the supply air to the space to change, and velocities to increase, possibly causing noise and overcooling or under cooling the space before the thermostat is able to compensate. This change in supply air pressure could be caused by the opening and closing of VAV boxes in the system causing severe airflow fluctuation.

This pressure-dependant control is rarely used as it has caused so many problems in the larger systems. In some smaller systems, where noise, accuracy, ventilation effectiveness, and comfort are not as important, it is still used occasionally. The first cost is typically much more economical than pressure-independent systems.

To overcome this problem, pressure independent controls were developed. These controls use two cascading control loops, meaning the control loops are chained together with the output of one establishing the set point of the other. The first loop (the space thermostat) controls space temperature. The output from this loop is fed to the second controller as the reset signal, setting the airflow set point required to cool the space. The set point range can be limited to both a minimum (cool and heat) and maximum airflow rate. The second controller modulates the VAV damper to maintain the air volume at this set point.

Air volume, velocity times its fixed duct area (CFM), is measured using a velocity pressure averaging and amplifying sensor that often has fairly sophisticated configurations (such as rings or crosses), so that accurate measurements of air velocity can be made even if inlet duct configurations are not ideal and even when flows are low. (See the ASHRAE Handbooks for more information on this subject.) Its operation is 'independent' of the changing supply duct pressure, although it still requires an acceptable range of duct static pressure in order to operate normally, usually 1–2 in of water column pressure.

The availability of the maximum volume limit is very useful during system balancing. The VAV box can be set to maintain the design air flow rate to the zone while the balancer adjusts manual volume dampers to achieve the desired distribution of air among the rooms served by the zone. The maximum volume limiter is also useful in preventing excess supply that can be noisy and objectionable. This might occur during early morning operation when spaces are warm and require greater than design airflow to cool down.

The capability to limit minimum volume is important to ensure that adequate airflow is provided to maintain minimum ventilation rates for acceptable indoor air quality. The thermal loads in the space (which dictate the amount of air supplied by the VAV box) are not always in synch with the ventilation requirements, which are a function of occupancy and the emission rates of people, pollutants from furnishings and office equipment.

The minimum volume control can be used to ensure that sufficient supply air is provided regardless of thermal load; this also promotes and maintains proper building pressurization as long as the exhaust and ventilation operation (and volumes) are coordinated into the control system as well. Pressure-independent box controls are preferred over pressure-dependant for many reasons, including both minimum and maximum flow rates can be controlled regardless of duct pressure, and flow rates are a function of available duct pressure.

Minimum volume control is also essential for reheat boxes (where used to provide space temperature or heating), which must be able to maintain a minimum flow and temperature to effectively heat the space. Reheat boxes behave like cooling-only boxes when the space is warm. As the space cools, the air volume slows to its minimum amount, and while maintaining minimum volume the air is reheated.

The reheat box function is also used to heat the cold air flowing into the room when the room thermostat is calling for heat. The volume cannot be too low or the air supplied will be too warm and buoyant, and will not mix well with the air in the space, resulting in discomfort and possibly inadequate ventilation in the occupied zone.

Remember the supply air from the VAV fan system is cool so, before any useful heating can take place, the air must first be reheated up to the space temperature to provide heating. Some VAV box controllers have the ability to control to dual minimum set points: one during cooling operation to maintain minimum ventilation rates, and a higher set point during peak heating operation.

Some control systems allow the supply air to be reset based on the return air from the spaces so that energy can be conserved during intermediate temperature and low-humidity conditions. Other triggers include a change in the heating water temperature, or outside air temperature. Some VAV controls have a discharge air temperature controller/sensor at its exit so that under-cooling and over-heating can be avoided. These discharge temperatures can also be used to reset the discharge air temperature control back at the AHU, and are excellent information for troubleshooting temperature complaints of the occupants.

Typical Constant Air Volume System, with Variable Speed Fan for Filter Loading

A variation of the VAV delivery system described in the last few pages is the constant air volume system that uses variable speed fans to account for increasing pressure drop across the filters as they load up with dirt. This form of control is used in healthcare buildings, laboratories and clean rooms, as the amount of air delivered to the space must be constant, so that precise building pressurization can be maintained. The air delivery to the space may still be thru VAV boxes, but the VAV boxes are used as constant air volume (CAV) boxes, with the maximum and minimum volumes set at the same value. The supply air feeding the boxes is again typically 55°F. In some operating suites, the supply air can be as low as 48°F. Normally, the CAV box will have some type of reheat: either hot water or electric. The thermostat and/or humidistat for the room will control the reheat, and maintain its room temperature set point.

Care should be taken to make sure there is enough reheat energy available at each CAV box to perform its job, and that the supply air temperature does not get too cold for the application. In retrofit jobs, a common mistake is to reuse existing reheat devices that have inadequate capacity to supply the proper amount of reheat. This is especially true if a new, colder, supply air temperature is being utilized. The CAV box controller has its minimum and maximum set point as the same value, so it modulates in order to maintain its airflow set point.

Back at the air handler unit, the variable speed drives, or vanes, on the supply fan modulate to maintain the static pressure set point in the supply duct. As the filters load up their resistance increases, the VFDs modulate the fan speed higher to overcome the added resistance of the filter, maintaining a constant static pressure in the ductwork, as sensed by its supply duct static pressure sensor, typically 2/3 to 3/4 of the distance from the fan to the end of the duct.

Now, over to the return and outside air ducts. In this example the exhaust fans are separate and they are in the space operating stand-alone, interlocked to the AHU, and constant speed. (Note that if this exhaust fan is serving washrooms Using *Figure 9-9* as a guide drawing, there needs to be a specific amount of outside air flowing into the AHU while the unit is on. There are many methods that good designers can come up with to achieve this (Chen & Demster 1996)).

First, measure and monitor the outside airflow into the AHU. Install an airflow-measuring device in the outside air ductwork to measure this flow. Flow is the (average) speed of the air times its (flow) area, so there needs to be a high enough velocity of air so that the velocity pressure sensor is accurate, and locate the sensor in a place where there is good nonturbulent flow. Place the VFD and fan in the return duct and put a static pressure sensor into the mixed air plenum. With the all return and outside duct dampers open, and the supply fan running at its supply duct pressure set point, the outside air flow is tested, adjusted and balanced into the unit using the return fan VSD modulating to a mixed air plenum static pressure.

The engineering economics concept here is clear; with all of the dampers open, the fan is not wasting energy trying to overcome damper pressure drop; and therefore there are energy savings with the use of the VSD over the traditional damper modulation. Now, once the proper ventilation outside airflow is reached, that mixed air plenum pressure becomes the set point for the return fan VSD to maintain, and the supply fan VSD modulates to maintain its set point and that will change the mixed plenum pressure and cause the

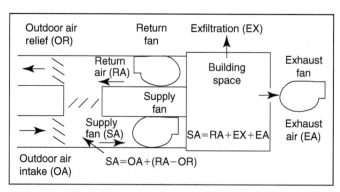

Figure 9-9 Schematic showing SA, RA, and EA Relationships

return fan to track it. There is an OA monitor in the duct that can monitor and record the measured OA flow, and its damper could be made automatic over a small range in order to make small adjustments. If further monitoring of OA effectiveness is desired, a CO_2 sensor in the return air duct can be used to monitor its value and record its trend. For operating suites and clean rooms, it may be desirable to have room pressurization sensors in the space as well, in order to monitor pressurization effectiveness.

Chiller Plant, Pumps, and Boilers – Monitoring and Control

Chiller and boiler plant control systems consist of controllers, sensors, relays, transducers, valves, and dampers that operate the plant and its equipment, as well as optimize it for energy conservation, efficiency, and functionality.

On a demand for cooling, the chilled water system shall be enabled. First the chillers respective chilled water primary pump may be energized and then the chillers respective condenser water pump (or condenser fan system) may be energized, then, several minutes later, after flow is proven at the pumps/fans/chiller, the respective chiller shall be enabled to operate and produce chilled water to the HVAC&R system. The internal controls of the chiller shall operate each chiller to produce the required chilled water supply temperature. The control system shall monitor the common supply and return temperatures and provide a constant common supply water temperature, typically 40°F (adjustable).

If the chiller plant has cooling towers as a source of condenser heat rejection, then the condenser water supply temperature shall be controlled to optimize the plant energy use (check with chiller manufacturer for recommended set point values and allowable drop in temperature at times of low enthalpy outside air) by staging the cooling towers as necessary and cycling the cooling tower fans as required. If cooling tower fans are controlled using variable speed drives, then their speed can be varied in order to meet the demand. Deadbands and time limits shall be set up such that the fans do not cycle unnecessarily.

Some cooling tower plants have a bypass loop and internal sumps that allows bypassing of water around the towers in order to maintain minimum set points during cold and below freezing ambient conditions.

On a call for heating, the plant control system shall start the boiler primary pump, and then the boiler shall be started–stopped. The boiler primary pump shall be operated at least 10 min before and after boiler is energized. A high limit temperature sensor will operate the circulating pumps until the boiler internal temperature is at or below its set point. The boiler discharge temperature control is integral to the boiler and is provided by the manufacturer. If multiple boilers are used, a control routine that starts and stops them is used to maintain an adjustable set point temperature of water in the loop or storage tank while maintaining an appropriate return water temperature. In addition, some boiler plant controllers monitor outside temperature, and provide higher hot water supply temperature set points to be available during extreme cold ambient conditions.

Primary pumping systems are used frequently and are described above. In some designs, secondary pumps are used. Secondary pumps receive their water flow from the primary loop usually via a coupling "header" as shown in Figure 9-10.

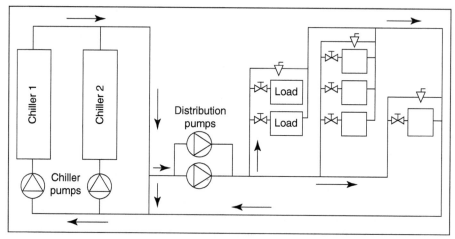

Figure 9-10   Primary Secondary Pumping System

The primary loop (chillers and pumps) provides the "header" with a constant flow of chilled water. The secondary loop, via its secondary distribution pumping system, takes water from the primary loop as needed for the hydronic systems it serves. The secondary pumps may be controlled in order to provide adequate flow to the secondary piping loop, using a water pressure or flow sensor. They can be constant speed pumps with a bypass arrangement, or the motor speed can be controlled by a variable speed drive, less energy waste. Control valves are typically two-way when variable flow secondary control systems are used.

Secondary pumps, as with any dual equipment scenarios, shall be rotated from lead to lag typically on a daily basis. When a failure is sensed, the lag (or redundant backup) pump can automatically take over.

A flow meter and appropriate temperature sensors, in the secondary loop, can be used to monitor its flow readings and calculate energy usage. Flow meters can be used to monitor make-up water flows in order to trend water usage and identify potential leaks.

When the outside air temperature is above/below an appropriate temperature (adjustable), the chiller/boiler plant may be disabled/enabled. An override feature that will override this outside temperature shutdown should be installed and available.

For maintenance or out-of-service times, the control system will sense the abnormal off and/or alarm condition of a chiller, boiler, fan or pump and bring on its next in series in order to maintain its set points. All alarms should be monitored by an alarm providing visible or audible alarm.

Temperature and Humidity Monitoring and Control

The use of temperature/humidity monitoring and control devices is very important in today's design of HVAC systems. Whenever possible, the use of both is desirable. When the humidity drops too low in the space, a humidifier is switched on in order to add a spray of steam, or atomized water, into the airstream in order to add water vapor and increase the relative humidity.

During periods of high humidity, the control system typically activates the cooling systems in order to cool the airstream below its dewpoint temperature so that the water vapor will condense on the cooling coil and be removed by the drain pan. Further dehumidification and adjustment of the relative humidity downwards can be obtained by reheating the cold airstream with some acceptable heat source, such as hot gas reheat, recovered heat, heat load of the building, lights, energy wheel, face and bypass, etc. (see ASHRAE Standard 90 for more information).

It is advisable to limit cycling of the cooling coil. Too much cycling of the cooling coil will not allow enough time for the condensate to drain from the coil, resulting in the condensate entraining itself back into the air stream of the supply ductwork, causing energy waste and potentially, IAQ problems.

Continuous monitoring, and recording, of temperature and humidity can be very useful when it comes to resolving problems and detecting poor performance.

Carbon Dioxide Control

In ordinary outside air, it is generally found that the carbon dioxide, CO_2, concentration is 350–450 parts per million (ppm), this is called the "background" CO_2 level. In city areas with high traffic volumes the level may be substantially higher. People breathe, inhaling air and absorbing oxygen and exhaling CO_2. The rate of carbon dioxide production depends on activity level and is quite consistent in the population. The higher the activity level the greater the CO_2 production.

A person in an enclosed space with no ventilation would continuously add to the CO_2 level. If ventilation is provided then, after a while, there will be a balance:

Background level in > Addition from person to raise concentration > Higher Concentration out

The process is shown in *Figure 9-11*. Note that the process is the same for one or many people with more than one person producing proportionally more CO_2 and needing proportionally more ventilation. For the ASHRAE Standard 62.1, default ventilation rate in an office is 17 cfm. The increase in concentration is about 620 ppm. If the background level was 350 ppm then the steady concentration in the office would be $350 + 620 = 970$ ppm. In the

Figure 9-11 Ventilation Air Collecting CO_2 from Occupants

case of a fully occupied auditorium, the ASHRAE Standard 62.1 default ventilation rate is 5 cfm and the increase in concentration is 2100 ppm providing a steady state level of $350 + 2100 = 2450$ ppm.

Note that the added concentration increases in proportion to the number of people for a fixed ventilation rate. Thus halving the population density in an office will result in the added CO_2 concentration dropping from +620 ppm to +310 ppm. Similarly, an increase in ventilation rate for the same occupancy will proportionally drop the added CO_2. For example, increasing the office rate from 17 cfm to 20 cfm per person will drop the rate to $+ 620 \times 17/20 = +527$ ppm.

This increase in CO_2 concentration can be used to control the ventilation rate. In the high population density auditorium situation, the level of CO_2 can be used as a surrogate (equivalent to) for occupant numbers and the ventilation rate adjusted to maintain a maximum CO_2 concentration. In the lower density office situation Standard 62.1 requires a ventilation rate significantly based on area per occupant and using CO_2 is more complex than we are covering in this course.

More detailed discussions and requirements can be found in ASHRAE Standard 62-2004 and ASTM Standard D6245.

Exhaust Fan Control

Exhaust fan controls are often thermostats or interlock relays specified with little thought as to when they are needed to operate. A typical example is the electrical room provided with a manually set thermostat with a range of 60–80°F. It gets set at 60°F and the fan runs continuously for the life of the building wasting energy.

Just as for the main system, you should be establishing what service is required. This may mean simply linking the exhaust fan in with the main system. However, for any system that operates out off occupied hours to maintain temperature or humidity the exhaust fan operation is probably not required.

The designer should provide a schedule showing each exhaust fan, its specified design parameters, and its intended control and interlock methodology. Care should be taken to ensure that these exhaust fans do not operate when the main HVAC systems are off, unless needed, as this can cause negative pressurization and cause unconditioned outside air to enter the building and cause IAQ problems. Where used for conditioning equipment rooms, a room thermostat should be used to turn the fan on and off to maintain its exposed temperature set point. In storage rooms, where sensitive materials are being stored, a thermostat and humidistat may be used in order to operate the exhaust fans and supplemental dehumidification/humidification equipment to maintain the space temperature and humidity requirements.

Toilet exhaust fans are usually interlocked to their respective HVAC system and its OA damper operation; sometimes the toilet exhaust is supplementally interlocked to a light switch. In garages and spaces where hazardous or toxic gases/fumes can gather, exhaust fans are operated continuously, or can be controlled by a high limit gas/fume-sensor/controller. For chiller plants where refrigerants are used, refrigerant gas monitors are set up to operate exhaust fans and/or open dampers/doors when their set points are exceeded.

In boiler rooms, exhaust and supply fans can be used in conjunction for exhaust of fumes and to also provide combustion air for the boilers.

For medical isolation rooms where harmful bacteria or germs etc., may exist, exhaust fan systems, coordinated with supply fan systems, are set up to maintain their individual room pressurization set points, either negative or positive, and should have a room pressurization monitor and alarm panel installed in each room. Visual and audible alarms should be initiated when set points are violated, and when the fans fail. In a lot of cases, depending on what the intended use of the room is, these exhaust fans may not be switched off for any reason, including fire alarm initiation. Isolation room fan sizing and controls should allow for controlling the pressure in the isolation room in situations when the ambient pressure changes, such as during strong storms such as hurricanes or tornados.

For flammable storage and/or explosion-proof areas, where the fumes are typically more dangerous and flammable than normal building exhaust, they are typically controlled using an "explosion-proof" or "nonspark producing" control device. The controls are typically constantly working and respond automatically to their temperature and gas/fume concentration set points by staging the exhaust fans on and off. In addition, visual and audible alarms should be initiated when crucial or safety set points are exceeded, and/or when the fans fail.

In almost all cases, exhaust fans need to be specified to be interlocked with the building fire alarm system to meet all local, state and national codes. Typically, all fans are stopped when the fire alarm system is in alarm, with the exception of engineered smoke control fans and some isolation room fans.

Fume Hood Control

Fume hoods are typically used in laboratories to collect and exhaust fumes. The cross-section of a hood in *Figure 9-12* has a vertical sash which can be raised and lowered as necessary. The horizontal air velocity into the hood is, typically, maintained at a constant velocity of 80–100 fpm whatever the sash height.

When the hoods are not operating, the room control is much the same as a normal occupancy for a variable air volume controlled room. When the hoods are operating, the control system senses the flows and pressures in the room, hoods, and ducts in and out, and makes adjustments to maintain the temperature, humidity, and pressurization of the room. The objectives for control in laboratory hoods are for the capture and containment of fumes and harmful gases, to maintain acceptable room pressurization, to maintain acceptable temperature and humidity set points, and to ventilate the space in order to preserve dilution of contaminants. Many conditions can affect performance and speed of response from the control system. There are great cost differences between desired levels of control for these systems; the designer needs to know and account for the accuracy and reliability of controls required for the process.

Variable air volume air handling systems are often employed in modern laboratory and hood systems. A fast and stable control system for the hoods, exhaust fans, supply fans, and all HVAC equipment is suggested. The airflows required are determined by the highest demand of the minimum

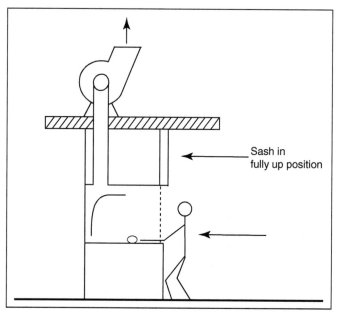

Figure 9-12   Rising Sash Fume Hood

ventilation rates prescribed, the cooling or heating loads required, and the total amount of exhaust from the hoods themselves. The airflow in and out of the room is controlled to meet these needs. Sometimes, the airflow required is below the amount needed to cool or heat the room to its comfort settings; therefore, the room supply air must be increased to meet the load. The lab control system must react to exhaust additional air through a "general exhaust" damper or air valve to compensate and maintain room pressurization and hood equilibrium.

Typically, the room is kept at a slight negative with respect to its adjacent corridors. Controlling the total supply air to be a little less than the total exhaust is what performs this. There are, however, situations where a positive pressure laboratory is required based on the type of experimentation being done.

Now as the hood sash opens and closes, the hood's exhaust volumetric flow is varied as a linear function of the sash opening percentage (which is an indication of the hood face velocity). Typically, there is a minimum flow that takes precedence when the sash is below 20% open. This relationship between the maximum and minimum flows is called the turndown ratio; for example if the maximum flow is 500 cfm and the minimum is 100 cfm, and the supply air is offset by 100 cfm, then the turndown ration required is 400:100 cfm, or 4:1. This maintains a constant face velocity into the opening of the hood. A constant face velocity is a key element in successful hood control. Typical set points of acceptable face velocities range from 60 to 120 fpm. A maximum of plus/minus 5% set point control should be maintained at all times. Unstable or excess velocities can cause adverse effects, such as loss of containment, turbulence, eddy currents, glass breakage, blow out of heat candles, and leakage from the hood. The control system should have less than 1s response timings.

There has been much research on control design and implementation. Many other factors such as occupancy in the fume hood areas, usage of the hood, cross drafts, sash positions, operator position, diversity of conditions, and placement of instrumentation, as well as temperature and humidity have a dramatic effect on some aspects of its ultimate control need for sophistication.

Condensate Management and Control

One source of microbial growth is improperly installed and nonfunctional condensate removal from cooling coil drip pans. Controls include drain pan float switches that shut down the fans and/or cooling systems when they sense an overfull drain pan. There are pipe sensors that fit in the condensate piping that sense a 'too wet' or 'too dry' condition, and stop their systems as well as provide alarming that can provide early detection of problems. To detect and alarm carryover in coils, a moisture sensor can be placed in the ductwork downstream of the cooling coil or humidifier to detect its alarm condition. Provide monitoring of the drain pan overflow and alarm the controller when tripped. To stop the condensate, stop the fan, or stop the DX cooling or close the cooling valve. If available, provide a sensor in the condensate pipe exit point so that it can sense the "lack" of moisture removal, which can predict that there is a problem with coil operation or at the drain pan.

Ventilation Monitoring and Control

The accurate and proper monitoring and control of ventilation into the system and into the occupied spaces of the building is very important and at the center of today's most urgent issues in HVAC controls and instrumentation. There are many ways to control the outside air entering the system, and the most successful is the direct monitoring with an airflow-monitoring device. A velocity-sensing probe is installed in the outside air duct. Depending on the size, and arrangement, of the duct, air-straightening vanes may be necessary. Follow the manufacturers' recommendations on mounting, but in general they need to be installed in a piece of duct with at least five diameters of straight length. The more tubes used, the more accurate and dependable the measurement. In general, one sensing tube per 2 ft^2 of duct is a rule of thumb. The velocity needs to be sufficient to create a readable and reliable differential pressure across the velocity tube, so in many cases the OA duct is downsized. This velocity tube reading is then multiplied by the area of the duct to get airflow in cubic feet per minute (cfm). The designer sets the amount of ventilation airflow in accordance with the local Code, current ASHRAE Standard 62.1, experience, expertise, and her/his best judgment. The airflow is monitored and controlled to this set point by modulating dampers, a fan, a variable speed fan motor, vanes, etc. Some means of documentation should be kept of these ventilation airflows in a permanent record.

Now that the proper amount of ventilation is being delivered to the supply fan, it needs to be delivered to the spaces in a proportional, appropriate and effective manner.

Filtration Monitoring and Control

Digital or analog differential pressure sensors, mounted across each filter bank, should monitor filter loading. When the differential pressure high limit is exceeded, a visual and audible alarm should be initiated. On medical systems, duct mounted visual exposed "Magnehelic™" differential pressure gauges, visible, with set point markings may be required. In some cases where the filters are concealed, a visible alarm light is required to signal dirty and ineffective filters. The manufacturer should provide the maximum dirty filter resistance and the alarm set point should be specified. For example, the filter resistance alarm set point might be 80% of the manufacturers' stated final resistance.

Outside Air Monitoring and Control

In some areas, the quality of the outside air is sometimes not acceptable. Sensors that can detect levels of gases such as carbon monoxide, sulfur dioxide, ammonia, acetylene, methane, propane, benzene, formaldehyde, hydrocarbons, HFC-CFCs, etc., are employed and installed in the intakes, ducts and spaces to provide protection against the introduction of sources of contamination such as garbage containers, gas sources, sewage, bird nesting areas, standing water, pesticides, cooling towers, etc. Typically, if the outside air is not acceptable, the control system can sense it with sensor devices, provide visual and audible alarms, and shut down the ventilation control fans and dampers for that period of time. You should consult the latest edition and addenda for ASHRAE Standard 62.1, Ventilation for Acceptable Indoor Air Quality, for specific requirements.

DX – Direct Expansion Systems

Some areas use DX (direct expansion) HVAC systems in their buildings. These systems use refrigerants and reciprocating, scroll and/or screw compressors to produce the cooling. The controls for these DX systems consist of manufacturer-provided unit controls which are installed and wired at the factory to operate the internal components, and there are building control devices that are field installed and wired to control the area of the building for space temperature and humidity. Sometimes they are all packaged into one large piece of equipment, called "package units" or "through the wall units", or sometimes the fan and airside equipment is separate from the compressor and condenser equipment, and this is called a "split system".

Control of the DX equipment is very similar to any other unit, but there is a potential for much colder coil temperatures since the refrigerants being used are being "directly" exposed to the airstream at the cooling coil. These refrigerants are interacting with the airstreams at very low temperatures, as low as 0–20°F. This means that the speed of response is much faster than for chilled water, and the potential for "freezing up" the coil is great.

First, the designer must establish an almost constant airflow and load across that cooling coil. It is not advisable to have variable flow across DX coils, unless approved by the manufacturer, as this could cause freezing. If reduced airflow to the space is needed, a bypass duct is used and supply air is routed to the return so that the airflow across the coil is constant. As the load changes, unloading devices and control schemes are used that have a sufficiently low minimum capacity. This will allow the unit controls to match and control to the changing load conditions.

When possible, specify a form of recoverable or waste heat for the unit, such as hot gas bypass/reheat, so that the controls can adjust the temperature and relative humidity of the leaving air stream. Sometimes this hot gas is also used to heat the potable water for a building.

Controls can be mounted in the control panel on the wall of the mechanical room or inside the units. The controller should control the stages of DX cooling in order to maintain the discharge air temperature at set point, using the stages of cooling and hot gas bypass/reheat where possible. The response of the DX cooling effect is very fast, and unloading and multiple staging controls is necessary to effectively stage the effect. The staging should also have a time delay between starts, so once a stage is turned completely off, it is typically not turned back on for 3–5 min. The supply air temperature set point that we are staging to maintain can have some resets embedded inside it, such as being adjusted by the return or outside air temperature.

For cold climates, a heating coil in the pre-heat position may be specified to protect the chilled water coils from freezing in the winter. Alternatively, the coils must be drained and dried or filled with antifreeze. For hot and humid climates, a reheating coil after the cooling coil may be required to provide reheat after the reduced air delivery required to produce dehumidification effects.

Water Source Heat Pumps

A very popular and effective design involves water source heat pumps. A supply of water is continuously pumped around the building with DX units connected to it, as shown in *Figure 9-13*. Each DX unit has the ability to pump heat out of the water and into the zone or to cool the space and reject heat to the circulating water. The system is particularly attractive in buildings with high internal cooling loads and perimeter heating loads. The system enables the excess heat in some areas to be used in cool areas needing heating, load transfer. The DX units operate as above, but their source of energy for heat and cool rejection is water, not the ambient air. The efficiency and capacity of the heat pump is greatly enhanced since it is not restricted by extreme ambient conditions. This circulating water source is typically maintained between 60 and 90°F year round, depending on the environmental and design conditions.

The pipe loop is typically connected to a boiler for heat when needed and a heat rejection evaporative cooling unit or chiller. The source of water for heating and cooling may also come from a lake, or a geothermal well/bore and pump system. Certainly, proper water treatment control is required for each of these scenarios. The evaporative cooler/cooling tower fans are controlled to maintain the water temperature set point as required. Variable speed fans are frequently used to save energy and improve control performance by modulating the speed of the cooling tower fans to maintain the condenser water temperature set point of the condenser water. A tower bypass valve can also be modulated in order to bypass water around the tower to mix return water with its supply water to maintain the condenser water temperature set point.

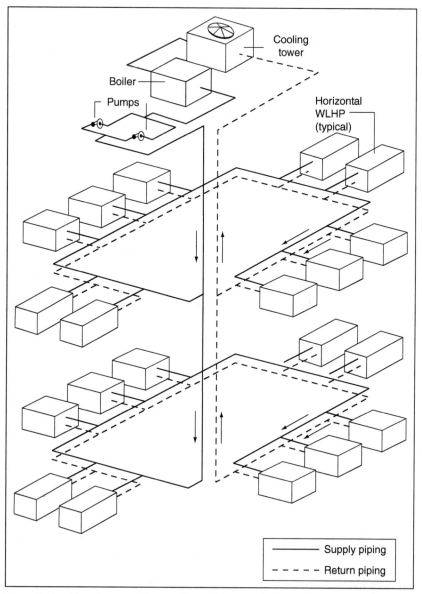

Figure 9-13   Heat Recovery System Using Water-to-Air Heat Pumps in a Closed Loop

In mild climates, a tower sump water heater may be cycled in order to tem-per the condenser water during cold ambient conditions to prevent basin freezing. In colder climates the cooling tower drains to an indoor sump avoid-ing the heater energy use and reduced reliability of depending on a heater to prevent freezeup.

Each heat pump unit can have a water valve that opens and closes upon operation of the unit. The water valve should be normally open and fail-safe (open to the coil) if possible. If a water valve is not used, then the water flows through the coil at all times. The pumping system is typically variable speed and provides enough water to the heat pumps as required by a water pressure sensor and control algorithm. If there is variable flow, then a two-way control valve should be used (if variable flow is not used, then three-way valves are required). A water pressure sensor at/near the end of the piping loop is provided to control the pump and maintain its pressure set point.

If the loop water temperature rises above or below its alarm set point, a visible and audible alarm shall be initiated, and the heat pumps affected should be stopped until the alarm is resolved. Back up and redundant equipment is suggested to be available to control any major parts of the HVAC&R system, such as pumps, tower fans, heat exchangers, back-up electric heat, etc.

Bibliography

Chen, Steve, Demster, Stanley (1996) *Variable Air Volume Systems for Environmental Quality*. New York: McGraw-Hill (especially Chapter 5).

Harriman, Lew, Brundrett, Geoffrey W., Kittler, Reinhold (2001) *Humidity Control Design Guide*. Atlanta, GA: ASHRAE (especially Chapter 2).

Persily, Andy (2002) Fall Issue of *IAQ Applications*. Atlanta, GA: ASHRAE (pp. 8–10).

Diamond, Mark (2002) Fall Issue of *IAQ Applications*. Atlanta, GA: ASHRAE (p. 1).

Montgomery, Ross D. (1998) *ASHRAE Journal*. July. Atlanta, GA: ASHRAE (p. 52).

The Holmes Agency, 2001, Mr. Raymond E. Patenaude, P.E., CIAQP. "Proper Controls for HVAC&R for Indoor Air Quality", St. Petersburg, FL.

ASTM D6245-98(2002) Standard Guide for Using Indoor Carbon Dioxide Concentrations to Evaluate Indoor Air Quality and Ventilation.

ASHRAE Standard 134-2005 Graphic Symbols for Heating, Ventilating, Air-Conditioning, and Refrigerating Systems.

ASHRAE Guideline 13-2000 Specifying Direct Digital Control Systems.

DDC Introduction to Hardware and Software

Contents of Chapter 10

Study Objectives of Chapter 10

This is the first of three chapters on DDC. DDC stands for direct digital controls. It introduces you to the basic hardware and software required to operate DDC controllers and their operator interfaces. The interconnection of these controls is covered in the next chapter while specifications, installation, and commissioning are covered in the final chapter. After studying this chapter, you should be able to:

Understand the types and performance of physical input and output points in DDC systems.
Have an understanding of the range of possibilities in DDC controllers.
Know the types of application software available and some relative merits of each.
Be aware of the capabilities available in operator workstations.

10.1 Introduction, and Input and Output Points

DDC are controls operated by digital microprocessors. 'Digital' means that they operate on a series of pulses, as does the typical PC. In the DDC system, all the inputs and outputs remain; however, they are not processed in the controllers, but all control logic is carried out in a computer, based on instructions called the "control logic."

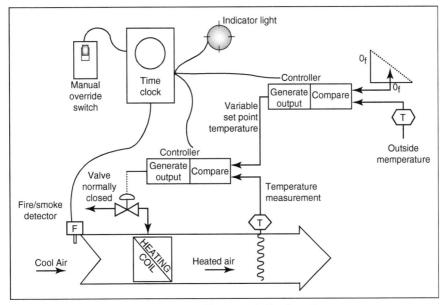

Figure 10-1   Air Heater with Outdoor Reset

Figure 10-1 is a simple control diagram of an air heater controlled by out-door reset. Power is supplied to the controls when either the timeclock or manual on switch requires it to run as long as the fire/smoke detector has not tripped.

The diagram in *Figure 10-1* is easy to understand, as most of the connections can be traced.

If this same control requirement was processed in DDC, then all the control functions and logic would be executed in software, run on a computer. If we were to take the same control diagram and use DDC to operate it, the control diagram would look like *Figure 10-2*. We have exactly the same inputs and outputs but they disappear into the DDC cloud. This chapter is about what is going on in that cloud.

Before we go inside the cloud, let us spend a moment on the inputs and outputs. In *Figure 10-2*, each input to or output from the DDC computer has been identified as one of the following:

- **On/off input** – manual switch, fire/smoke detector
- **On/off output** – power to light
- **Variable input** – temperature from sensor
- **Variable output** – power to the valve.

These are the four main types of input and output in a control system. Let us consider each one briefly in terms of a DDC system.

- **On/off input**. A switch, relay, or another device closes, making a circuit complete. This on/off behavior has traditionally been called "digital." Therefore, in terms of DDC, it is generally called a **"digital input"** or **DI**.

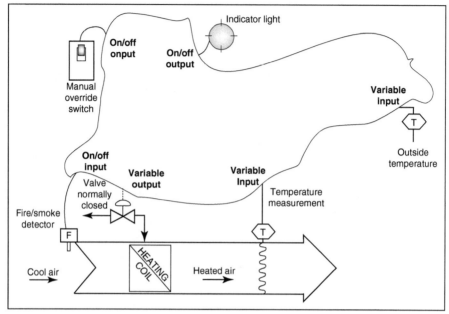

Figure 10-2   Air Heater I/O points

The term "digital" is not considered technically correct, since there is no series of pulses, just one "on" or "off." Thus, for on/off points the term "Binary" is considered more correct, and the term is being encouraged in place of "digital". So, "binary input," BI, is the officially approved designation of an "on/off" input.

Some changes in binary inputs need to be counted. For example, a power meter may switch for every kWh. If this is slow, a regular digital input may suffice, but often a high speed **pulse input** is required. This is a BI designed for rapid pulsing at speeds typically to 100 cycles per second, or more.

• **On/off output**. The on/off output either provides power or it does not. The lamp is either powered, "on," or not powered, "off". In a similar way, this is called a "**digital output**," **DO**, or, more correctly, "**binary output**," **BO**.

• **Variable input**. A varying signal, such as temperature, humidity or pressure, is called an "analog" signal. In terms of DDC, the input signal from an analog, or varying, signal is called an "**analog input**," or **AI**.

• **Variable output**. In the same way, the variable output to open or close a valve, to adjust a damper, or to change fan speeds, is an "**analog output**," **AO**.

You might think the next step is to connect these BI, BO, AI, and AO points to the computer. Things are not quite that simple. A sensor that measures temperature produces an analog, varying signal such as 0–10 V and our computer needs a digital signal. So between each AI device and the processor there is an "**A/D**," "**analog to digital**," device. These A/D devices, or AIs, are usually built in with the computer.

Having introduced you to point types and their designation let us now consider the contents and operation of our DDC unit.

The minimum components our DDC controller must contain are:

Power supply	– to power the computer board at a low dc voltage and the I/O points with dc and 24 V ac
Computer board	– to run the control software
I/O board	– with screw connections for wires in and out, A/D and D/A conversion hardware
Communications port	– a connection to enable the computer board to have thesoftware loaded and changed when required

The three huge differences between DDC controllers and all the other control systems we have considered are in

1. the computer board – all processing logic being done by microprocessors;
2. digital rather than true analog signals – raising issues of accuracy and timing;
3. the communications port – providing access for operators and other controllers to communicate with the controller.

In the computer, we can have software that can be as simple or complex as we wish, or can afford. The digital computer does everything in sequence, not continuously as it happens in, for example, a pneumatic system. The computer is fast but it still reads inputs discretely: one at a time. The communications port provides the opportunity for the controller behavior to be modified from a remote location and to interact with the operator and other controllers. For the controller in *Figure 10-2*, the set points can be adjusted from the plant operator's desk, and the unit could send a message to the boiler plant when it does, or does not, need hot water. This little system is shown in *Figure 10-3*. The controller communicates over a network cable to the operator and to the boiler control panel. The subject of networks and communication is covered in the next chapter and operator workstations at the end of this chapter.

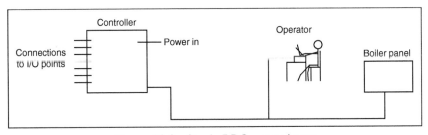

Figure 10-3 Simple DDC system layout

10.2 I/O Point Characteristics

Physical input and output points are made by wires connected by screwed connections to the I/O board. The simplest point is a BI. It can be internally powered, as in *Figure 10-4a*, or externally powered, as in *Figure 10-4b*. With the external power supply, the system has no way of knowing if the power supply has failed, or if the contact is open, so the internal power supply is more reliable.

Even with the internal power supply, a broken wire cannot be detected. With DDC there is a simple way of monitoring the circuit. *Figure 10-5* shows a monitored BI using an analog input point.

There are four possible situations:

- Short circuit, terminal to terminal voltage 0
- Switch closed, terminal to terminal voltage 5.0 (shorts out the bypass resistor)
- Switch open, terminal to terminal voltage 6.6 (the circuit is completed through the bypass resistor)
- Open circuit, terminal to terminal voltage 10.

The software can now be programmed to discriminate between these four situations: 0–4.5 V short circuit, above 4.5 and below 5.5 V switch closed, above 5.5 and below 6.1 V switch open, and above 7.1 V open circuit.

The analog to digital converter circuit is connected across the terminals and has a very high resistance, and so does not influence the basic circuit as drawn. The converter and the power supply must be adequately protected to allow for the inadvertent short circuit and for 24 V being applied to the terminals by mistake.

This ability to use an AI as a multiple BI can be used in many ways. Another example is monitoring a constant speed fan using a current ring

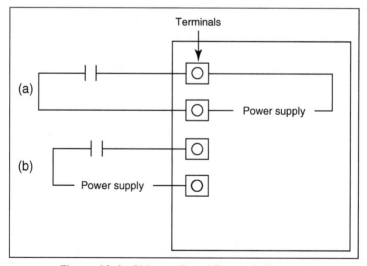

Figure 10-4   BI Internally and Externally Powered

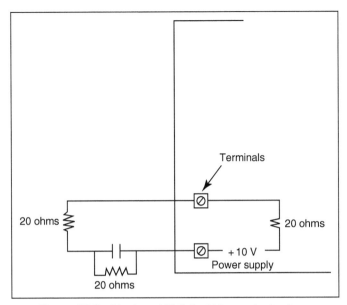

Figure 10-5   Monitored BI Point Using an Analog Input

around one of the supply cables. Typically, four situations can be detected on a belt drive fan: no current means the motor is not powered; low current means the motor is running but there is no load perhaps because belts have broken; around normal current, overloaded. This is an advance on simple proofing with a fixed current sensor, or sail switch, providing a BI input.

Note that the use of the AI has software costs not only in the control logic but also in the person-minutes to add the necessary operator alarm messages. Using an AI instead of a simple on/off BI costs more and the additional cost must be balanced against the added future value of the information on that particular fan. It is probably not worth it on a small fan but the fan that serves a whole building is a likely candidate. This possibility of using more complex and more costly control options with larger plant should be remembered and clearly addressed in any specification. Be aware that in the very common "low bid" situation the added future value will not be counted. If you know this to be the client mindset do not waste your time, or the bidder's time, putting costlier options in and then taking them out to minimize the first cost.

The analog converter is taking in a true analog signal and converting it to a digital signal: a series of 1 s and 0 s. The number of digits in the digital number determines the available accuracy. These are called binary numbers:

1 bit   0, or 1 - total 2
2 bits   00, 01, 10, 11 – total 4
4 bits   0000, 0001, 0010, 0011, 0100, 0101, 0110, 0111, 1000, 1001, 1010, 1011, 1100, 1101, 1110, 1111 – total 16
8 bits   total 256
10 bits   total 1024
12 bits   total 4096

If the converter is an 8-bit converter it can only pass on a change of $1/256$ of the incoming signal. For a cool climate, the outside temperature sensor range may be $-40°F$ to $+120°F$, a range of $160°F$. The converter would thus provide a digital signal changing every $160/256 = 0.625°F$. This could be quite adequate for controlling the economizer dampers based on temperature. However, if this signal is also used for calculating the enthalpy of the outside air for comparison with return air, the accuracy is inadequate.

Changing up to a 10-bit converter improves the discrimination to $160/1024 = 0.16°F$ which is probably close to the sensor accuracy for normal installations. Moving up to 12-bit $160/4096 = 0.04°F$ is certainly making the converter as accurate as the sensor. Note that although 12-bit numbers work well for control signals they have severe limitations for accommodating records of energy use. It is easy to need to store numbers in the millions when even 24-bit numbers (1,048,576) are inadequate for direct storage of accumulated consumption. There is a trick to dealing with this challenge. One sets up more than one register to store the number and when the first register is full one adds one to the second register and restarts the first register. For a 12-bit system the first register nominally stores numbers up to 4096 and the second register stores how many 4096 s have been accumulated. So if the first register is at 2134 and the second at 17 this means the count is $17 \times 4096 + 2134 = 71,766$.

Note that when you are defining accuracy in DDC systems there are five accuracies which combine to produce the "end-to-end" accuracy. They are

- medium to sensor
- sensor
- transmitter
- interconnections
- A/D converter.

Examples of medium to sensor accuracy being compromised are turbulence in the medium through a flow meter, outside air temperature sensors being influenced by solar radiation, poor mixing of air streams before a mixed air temperature sensor. In each case, the end-to-end accuracy can be severely compromised by this medium to sensor error. Then there is the sensor accuracy, what quality was required and what quality was provided? The transmitter performance must match the sensor and required range. A transmitter designed to provide 0–10 V with a thermistor over a range of $100°F$ is not going to be very accurate measuring chilled water flow temperature in the range $38–45°F$. If wiring is short, it should not significantly influence accuracy. For longer runs it may be better to use a 4–20 ma current loop rather than the more economical 0–5, or 0–10 V. Last is A/D conversion accuracy, which is discussed above. You are not expected to define each of these but rather to understand that these points of inaccuracy exist and to define a required end-to-end accuracy.

If accuracy is an issue, how can you asses the performance of a sensor system? By using a competent contractor who has equipment that has been calibrated by a National Institute of Standards and Technology (NIST) traceable calibration organization in the USA, the National Physical Laboratory

(NPL) in the UK, the Australian Commonwealth Scientific and Research Organization (CSIRO) in Australia or other similar national measurement standards organizations. Note that the HVAC industry serving commercial and institutional buildings has, historically, not focused on accuracy of measurement and control as the market has demanded adequacy of performance at lowest bid cost. If you need accuracy, be very selective in your choice of supply and installation companies.

Having converted the signal from analog to digital, it can then be processed before being used as an input to the process software. For example, the square root may be taken when the incoming signal is velocity pressure so as to provide a signal proportional to velocity. The signal can also be smoothed. In an air system VAV box, the measured velocity signal flutters around so it is common to smooth the signal. Let us suppose the system is checking the velocity signal every second and keeping a value called "old signal." Using a 5-s smoothing every cycle, the following calculation is made:

$$(\text{new signal} + 4 \times \text{old signal})/5 = \text{old signal}$$

The "old signal" is the smoothed signal that is used for the controller input.

A different input change is the change in set point. This may be due to changing from unoccupied mode to occupied mode with a corresponding change in set point or simply the occupant changing the set point. In this situation, it is common to include a ramp function. This ramp limits the rate at which the control variable can change. The controller is thus steadily driven in a direction rather than given a shock from which it has to recover.

Inputs can also be modified based on a lookup table or formula. Thermistors are non linear and many manufacturers standardize on a specific thermistor and provide AI points with the lookup table, or algorithm, built in. The lookup table or formula can also be used to deal with outdoor reset. The information that the water temperature is to be 70°F at outside temperature of 60°F and 190°F at −10°F is all the data needed to set up a lookup table or formula to convert from outside temperature to required water temperature. In other words, some mathematics in the software replaces the controller to provide the outdoor reset signal.

Now let us consider outputs AO and BO. The issue of conversion accuracy is similar on analog outputs as it was on inputs. The difference is that many driven devices, such as valve and damper actuators, do not react to minute changes in signal. They need a significant signal change to overcome friction; so 8 bits, giving a nominal 256 increments of change, each of 0.4%, is adequate even though 10-bit resolution is typically provided.

The output power of AO and BO signals is very limited so the signal is frequently used as a control signal not to provide control power. In the case of pneumatically driven valves and dampers the electric to pneumatic, E/P, transducer will typically be mounted close to the valve to minimize installation cost and maximize response speed. The transducer may be provided with adjustments but it is far better to make all adjustments at the DDC panel where changes are specific and can be recorded.

This section has briefly introduced you to the inputs and outputs from the system.

10.3 Control Sequences

Writing control sequences is neither easy nor quick. Following a clear and logical progression significantly increase is the speed and precision. The following is one step-by-step process. You might also note that large sheets of paper are much easier to get started on compared with the standard office notepad.

Step 1. Make a sketch of the system. Divide the system into subsystems.
    The typical systems we have been looking at include mixing dampers, cooling coils, heating coils, and fans.

Step 2. Choose each subsystem and identify what is to be controlled and the main process variable.
    In an air-handling unit, this might be supply temperature and cooling coil chilled-water valve, time and fan running, and duct static pressure and fan speed.

Step 3. Establish the control relationship between the process variable and the controlled equipment.
    In the case of the cooling coil, we can have a straight forward PI control using duct temperature input, our process variable, and output to the E/P for a pneumatic control valve.
    If this was an air system, the process variable could be velocity pressure in which case the process variable could first have been smoothed by taking a rolling average over 5–10 samples and the square root to get velocity and then multiplied by area to get volume. This volume would then be used as controller input. Note that multiplying by the area makes no difference to the controller operation, but it is much easier for the operator as a check on the value when it is in, understandable, useful units.
    The control method chosen is often based on the required speed of response, or time constant, of the control loop. Zone temperature control and outdoor reset are both fairly slow control loops. For these, a simple proportional band control with set with a narrow proportional band will often provide excellent service. For average speed loops such as economizer dampers, heating coils, and cooling coils the PI controller works well. As mentioned before adding the derivative, PID, makes tuning more difficult and often does not noticeably improve performance.

Step 4. Then list external factors which will affect each control loop.
    For our cooling coil these may include being interlocked so it cannot run when the heating coil is operating, may be disabled at night, will go on and off as the economizer takes on the cooling function, and may be delayed in opening if the chiller has just been started.
    For all the control loops there may be overall factors such as running when occupied, running when unoccupied maintaining occupied temperature, running when unoccupied to prevent excess humidity, running when unoccupied to maintain setback temperature, maintaining running with fire settings when fire alarm trips, running under Fire Department control.

Step 5. Establish where, and how, the factors in Step 4 are generated.
    This step is working out how the signals from Step 4 will be generated and where they will be available. For instance, a delay on opening the chilled-water valve when the chiller has just started will come from

the chiller. Since this factor will likely be used by all the chilled-water valve circuits the chiller should broadcast the information.

Step 6. Work up of the logic.

Using the data collected form the previous steps it is now possible to work up the control logic flow for each loop.

Step 7. Work out all the limits which apply, when they are to be applied, and how they are to be applied.

Limits may be absolute limits of temperature, flow, and pressure or limits on rates of change. For example, in an air-handling unit what is the minimum allowable supply temperature for an alarm, to shut the plant off. This step typically generates the requirements for safety and alarms. Thus the humidity sensor in a supply duct could provide a signal that was used to start overriding the controller when the duct humidity rises above 85% and sets of an alarm if it rises to 90% relative humidity.

Throughout this process it is important to be very clear about each component of the system. One way to do this is to consistently use a point naming convention. Many systems are based on a hierarchy of elements such as:

Type – System – Point – Detail

"Detail" allows for a number of identical points. On a larger system there may well be a smoke detector in both return duct and supply duct. They are physically the same but need to be differentiated in their name. Some points on an air system AHI might be:

AI AH1 OAT	Analog in, sensor, outside air temperature
AI AH1 ST	Analog in, sensor, supply air temperature
BO AH1 SF	Switch out, supply fan on/off
BI AH1 SF	Fan proof switch
AO AH1 SFS	Variable speed, supply fan speed controller
BI AH1 SM 1	Switch, smoke detector 1.

On a site with many buildings, a letter code (ideally an abbreviation) will usually be included before the system, e.g. AI DUF AH1 OAT. If buildings are on multiple sites, the coding may have a site code and then a building code. The critical issue is to be clear and consistent about your system. It can be a well worthwhile exercise trying it out on imaginary systems to make sure it works in all situations before having the contractor start detailed drawings. The naming has been introduced starting from real points such as analog and digital in and out points. You may wish to extend your convention into standard virtual points such as plant running hours.

If you write these point names on the system diagram and in your specification, it will be easier to keep track of what's what and where. It will also be easier to require the contractor to follow your naming requirements. Finally, have the point names on the items in the field. It is both frustrating and a waste of time for the maintenance staff faced with two identical electrical boxes and identical connections and wires inside and not knowing which is the temperature sensor and which is the humidity sensor.

10.4 Software Introduction

We have the IO points and control sequences ready now let us get down to considering some software routines to use them.

Let us consider economizer damper control and how the software logic is worked up. A very simple situation could have the following requirements:

1. When fan is off, damper to be closed (0% open).
2. When space occupied, and fan is on, damper is to open at least to minimum position of 20%.
3. The damper % opening is to adjust to provide 55°F mixed air when the temperature is below 67°F using proportional control with 4°F proportional band.

We can work up the required logic knowing that proportional control is linear and that, in this case, the damper will try to be open 0% at mixed air temperature (MAT) 53°F (55 − 4/2) and 100% at 57°F (55 + 4/2). Between those points the proportion open will be:

$$\frac{(\text{mixed air temperature} - 53)}{(57 - 53)} \times 100\% = 25 \times (\text{MAT} - 53)$$

We can work up the logic based on calculating and repeatedly storing the value of a variable called *"damper output."* Going through the logic, making changes to the variable, will provide a value of the damper output to finally be sent to the damper actuator.

Logic

1. Calculate *damper output* based on mixed air temperature (MAT): 25 × (MAT-53).
2. If outside air temperature above 67°F change *damper output* to zero.
3. Pass on highest of *damper output* and minimum damper setting (20%) as *damper output*.
4. If timeclock at unoccupied set *damper output* to zero.
5. If fan off set *damper output* to zero.

Does this work? Try it for yourself for a day when the outside temperature is 89°F, when the fan is on but the timeclock is at unoccupied. Now work out the situation for a day when it is only 40°F outside with the fan on and the timeclock is at occupied.

This example is extremely simple. If the logic required integral or enthalpy control the logic would have been difficult to work up. In addition the logic is the same for each similar situation. Manufacturers have thus worked up standard routines for the standard pieces of equipment such as dampers and for standard groups of equipment such as air-handling units, VAV boxes, and heat pumps. However, any variations from standard routines still have to have their logic worked up and implemented in the system software. The control software is called application software.

Application software falls into two categories: general purpose and application specific. General-purpose programming languages allow the control system to perform almost any control function and software may be custom-written for each application. Application-specific software, on the other hand, is preprogrammed for a specific, common HVAC application such as controlling a typical air handler. Commercially, application-specific controllers are available where a number of choices are made from a menu of hardware and software choices. For example, a very common application specific device is a residential thermostat. It may, as standard, include the ability to control one or two stages of heating, one or two stages of cooling, and show the outside temperature from a remote sensor. In a specific installation, this thermostat could be used with just one stage of heating and one stage of cooling. The extra functions are simply not used.

Three styles of application programming are commonly used for general-purpose languages: text, graphical, and ladder-logic.

- Text-based languages. Text-based programming languages resemble standard computer programming languages. In some early DDC systems, they resembled assembly language programming with four-letter, semi-mnemonic elementary functions, each having several numerical parameters that only a specially trained programmer could understand. These rather elementary languages have been replaced by variations on high level programming languages such as BASIC, FORTRAN, C, or Pascal. Special subroutines are usually provided for typical HVAC control applications, such as functions to turn equipment on and off with optional time-delays, set point reset tables, minimum/maximum value selection, psychrometric functions (such as determining dew point from dry bulb temperature and relative humidity), optimum start routines, and, most importantly, functions to perform PID control loops.

An example of a BASIC-like language is shown in *Figure 10-6*. The system will run this routine every so many seconds. Note the comment statements, C***, that assist the reader in understanding the logic of the program. The program ignores these comments. Also the line numbers have been chosen to separate the blocks of logic. Can you work out what is going on?

Line 110 checks if the fan is on. If it is on the program skips to line 510. If fan is off the program sets damper position to 0, closed, and skips to the end of the program. If the fan was on the double switch sets temporary value $LOC3 to 1 if the temperature is below 63. If temperature above 65 it sets $LOC3 to 0 acting as a thermostat with 2-degree dead band. Line 520 sends the program on to the mixed air damper control if $LOC3 is 1. Otherwise the program goes to line 530 and sets the outside air damper to 20% before skipping to the end of the program. Line 610 repetitively calculates the damper position, $LOC1, and line 620 chooses the maximum of $LOC1 and 20 as the required damper percentage open.

Although simple in concept writing, effective code is a learned skill. Not only must one know the exact structure of the code blocks such as IF, LOOP, MAX but also how they can be efficiently and effectively used.

```
100   C *** ECONOMIZER OUTDOOR AIR DAMPER CONTROL
105   C *** SHUT DAMPER WHEN FAN IS OFF
110   IF (FANDPS. EQ. ON) THEN GOTO 510
120   SET (0, OADMPR)
130   GOTO 1000
500   C *** SHUT OFF ECONOMIZER WHEN OAT > 65
510   DBSWIT (1, OATEMP, 63, 65, $LOC3)
520   IF ($LOC3) THEN GOTO 610
530   SET (20, OADMPR)
540   GOTO 1000
600   C *** OUTDOOR AIR DAMPER, N.C., 0 TO 100%
610   LOOP (128, MATEMP, $LOC1, 55, 10, 30, 0, 1, 50.0, 0, 100, 0)
620   MAX (OADMPR, $LOC1, 20)
1000  C *** END OF ECONOMIZER CONTROL
```

Figure 10-6   Text-Based Programming

Text-based programming should provide actions for all possible situations, be modular and structured, well documented, and be parameter-based.

Important factors in having a complete and robust installation include:

(a) *Structured code.* Most advanced high-level languages have developed a structured format; that is, the use of IF-THEN-ELSE blocks and the use of the GOTO statement. Most languages will allow both forms, and it is the skill and style of the programmer that select the form to be used. Most experienced programmers prefer the structured form. The advantage of the structured code is its clarity, allowing a person examining the code to find the relevant information and instructions to the system in one place in a readily understandable formation.

(b) *Documentation.* Many of the programming languages have a capability to insert English-language text (comments) into the program to describe the desired function of the section of code. A good programmer uses comments to make the program easy to read and understand. While many programmers avoid this as unnecessary because they understand what they did, it clearly aids troubleshooting for both the original programmer and others who are examining the system. After the job is installed and time has passed, the programming is no longer intuitively obvious to understand. Use comment statements liberally to enhance the lucidity of the program and the system. Putting in a comment when a change is made can also assist in problem resolution.

There are other factors to consider when selecting and programming a text-based system. Another issue concerns whether the code exists on-line or off-line. In some systems, all of the programming resides in the field control panel; in others, the program code exists in the operator workstation and only the "compiled" machine language exists in the field device. The latter method minimizes the memory requirements of the field device and also may increase the execution speed of the program by the processor. The program will define what conditions or modes will affect the element.

Component-style programs typically involve smaller and more numerous modules. It allows a user or programmer to quickly determine why a system element is doing what it is doing and when it should be doing it. It also allows the programmer to be thorough and robust, as described above, as he/she would know that all possibilities are covered. In some systems, the path through the software can be displayed so that it is easier to trace the logic being executed at the current time.

Typically, systems are controlled to maintain set point or target values. For example, the system attempts to maintain the room temperature at a set point value determined by the operator or occupants of the space. While it is easy to insert these values directly into the program code, it is usually desirable to set these up as variables so that system operators may reset them in the future. This will allow slight modification and tuning of the system to best match the building condition and occupants without the need to reprogram. While it is useful to code these constants into the program, do not bury the values in the code. It is often best to define the values as variables and set their value within the program but at only *one* location. A programmer/operator may easily alter their values in the future.

Much modern control system software allow for "menu-driven" platforms and configurations. This "fill-in-the-blank" menu software is much easier to use. Almost all of the programming is done for you, all you have to do is fill in some blanks for operational parameters such as set points, schedule times, PID parameters, on and off timers, existing conditions, etc.

- Graphical programming. Graphical programming is based on logic flow charts. Graphical programs are much easier to "read" and "comprehend" or "visualize" than text-based programs, as can be seen in *Figure 10-7*. Take a moment and work out the program logic in *Figure 10-7*. This is the logic:
  PID Control - if **Fan Status** is on use **Temp**, mixed air temperature, to calculate required damper position.

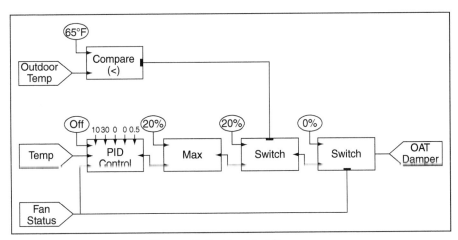

Figure 10-7 Graphical Program

Max - chooses the maximum of **PID control** output and 20%.

Switch passes the value from **Max** if **Compare** detects the outdoor temperature is below 65°F otherwise 20%.

Switch passes the value from first **Switch** if **Fan Status** is on, otherwise 0 to close the damper.

Graphical program languages can resemble block diagrams and pneumatic control diagrams, which improve their acceptability among some control system designers who may never have felt comfortable with computer-based controls. These programs are self-documenting, unlike text-based programs that usually also require flow charts to fully understand the logic they are implementing. As a normal rule, the software comes with a lot of typical examples and HVAC&R symbols already drawn for the user. These are generally configured as a common formatted graphical language such as bitmaps, gifs, or jpegs, and are changeable with common programs such as Paint, or CorelDraw, or PowerPoint. Graphical programming is expected to gradually replace text-based programs, much as Windows© has replaced MS–DOS© operating systems on many PCs. It is almost like a calculator where graphic symbols such as "+" and "−" represent a thing or action desired. These are sometimes referred to as "icons." However, many technical people and engineers still prefer the text language. One disadvantage of graphical programs is the need for a powerful PC/laptop with a high-resolution monitor and faster graphics adapters. Often the simple handheld or portable operator interfaces, which does not carry much memory or sophistication, still operate in the text-mode.

Graphic-based programming was developed in order to eliminate the need for system users and for the person setting up the system to know how to program in a high-level computer language. The user must still be computer literate and must understand what he or she wants the program to accomplish. Words such as "intuitive" or "obvious" are often used to describe the programming process by those who have been doing it for years! Do not imagine that it is intuitive or obvious to most people or that it can be learned in a day or two of manufacturer's instruction.

The process typically entails placing, dragging, and arranging "function blocks" on the computer screen and connecting these blocks with lines to indicate the relationship of control system data between the various blocks. A block may consist of a DDC point or element or may be a calculation, action, or logical evaluation. Shapes used include diamonds, squares, triangles, ovals, circles, parallelograms, etc. These more complex blocks usually will have a number of parameters that must be entered in order to tell the computer how to interpret the block and what to do with it.

A person who understands controls should be able to become proficient in graphic-based programming in a much shorter time than required to become proficient in a high-level programming language. The converse of this feature is that it often requires more time to program the system. Programming also may be slower where complex strategies or dense equipment drawings are required that are used or in large systems where not all interacting components of the system may be viewed on one screen and time is required to navigate the physical screen display to tie in the relevant components. One advantage of graphic-based programming is that a flowchart of the system's

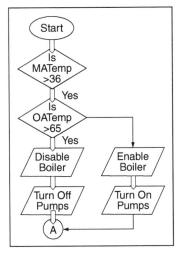

Figure 10-8   Flowcharting

sequences of operation is produced as part of the programming process (*Figure 10-8*).

After using these tools comes the finished product that is intuitive and easy to comprehend.

The usefulness of the flowchart often depends on the ability to display the entire set of building's sequences, as interaction between just a few of the components may be confusing or incomplete. With most control systems it is not difficult a task, as they can be broken down to their basic elements and the flow chart works out well. Graphical programming can even become "faster" than text-based programming once the user gets proficient; also, the use of duplicate graphics for multiple similar pieces of equipment and the ability to "copy" one graphic system to another and just change the hardware assignment can be helpful.

Ladder-logic DDC systems and programmable logic controllers (PLCs, commonly used in industrial controls) often use ladder logic for programming. Ladder logic is simply Boolean (relay) logic (see Chapter 2) implemented in software rather than using actual relays. Fairly sophisticated control sequences can be constructed with ladder logic programming, but not as easily as the other graphic techniques described above.

Some systems combine graphical and text-based programming. With these, the programmer constructs a graphical schematic of the system being controlled, then the programming software automatically writes text-based software typically used in the application selected. The programmer can usually edit this text-based software to refine the control logic to the specific application.

Other controllers use what is called logic control and interlock loops, which involve menu-driven programming of a combination of AND, NAND, OR, NOR, XOR digital gates. Also, analog gates are similar that allow comparison of analog values for the control system to pseudo or software set points (*Figure 10-9* and *Table 10-1*). The way the gate logic works is simple and basic.

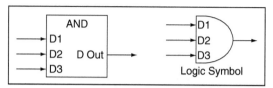

Figure 10-9   AND gate

Table 10-1   AND Logic

Statement # 1	Statement # 2	Result-and-Gate Response
True	True	True
True	False	False
False	True	False
False	False	False

For example, the "AND" and "OR" gates are the most popular and useful. The logic compares two or more true/false inputs and, based on its logic, produces an output of true or false. In the case of the "AND" gate and two statements for example, it asks the question if each of the statements are true, i.e. true and true? And the result of a true and a true is a true; otherwise it is a false result. For example, this "and" logic can be used in the following application. In an AHU control valve, control loop that produces a signal that modulates a control valve for cooling, there can be an "and" gate written that would say, if the AHU fan status is on, or true, and the OA damper is open, or true, and the AHU safety devices such as a smoke detector contact or drain pan float switch contact are normal, or true, then the "and" gate can be used to allow the operation of the control loop; if for some reason 1 or more of those inputs go false, then the "and" gate will cause the control loop to stop.

Similarly, for an "OR" gate, if any of the statements are true, i.e. a true and a false, a true and a true, or a false and a true, would result in a true result. If both statements were false, then the result would be false. In other words, a true or a true is a true, otherwise false.

This can be shown diagrammatically in many ways as seen below.

For the "AND-Gate" example, the definition is that if the stated fact #1 "and" stated fact #2 is true, then the AND gate response is true. All others result in a false response. (A "NAND" gate result is the opposite result of this.)

For an "OR-Gate" example, the definition is that if the stated fact #1 "or" stated fact #2 is true, then the OR gate response is true. All others result in a false response. (A "NOR" gate response is the opposite result of this.) For a special purpose example called a "XOR-Gate," the definition is that the result is true, if and only if, just 1 of the stated facts is true. These simple logic gates can be used in combination to perform interlocks and complex "relay logic" algorithms. Remember our convention is that if the result of the gate is true, then this disables the desired output, or "open's" the relay contact, and makes it go to its normal position.

For example, consider how to interlock the EF (exhaust fans) with the AHU. When the AHU is on, then the EF is commanded to start, and vice versa. What would be done is to configure a "NAND" gate, with the AHU status linked to each input of the NAND gate. Then when the status of the AHU is on, both of the status will be on or "true", the True and True in a NAND gate result in a False output (the opposite of an AND gate function). The false output would allow the EF to operate. When the AHU goes off, then the status goes to false, a false and a false results in a true output, and the EF point is disabled or stopped.

For a "NOR" gate in order to use it to announce alarms; for instance, if there were five alarm conditions that needed to set off an alarm light or horn. Put all of the alarm statuses as inputs to the NOR gate, and if any alarm is active, or true, or if all alarms were true, the result would be true; reversed for the NOR gate, the result would be false, which would not disable the alarm light or horn. If all of them are false, then the NOR gate result would be true, and disable the alarm output.

The analog gate works similar to the digital gates described in *Figure 10-10*. A value can be inputted and compared to another value such as outside air temperature. Mathematical operands such as plus, minus, equal to, etc., are used to compare. If the statement is true, then the output of the analog gate is true, and vice versa. An example would be to compare the mixed air temperature to a set point such as 32°F. If the MAT were less than 32°F, then the output would be true. This true gate statement could be used in an outside air damper open-close loop, and force the damper closed when the MAT is below 32°F.

With text-based and ladder logic programming, the source code (the actual text that the programmer wrote) can be stored in the DDC panel memory itself; or off-line. Off-line storage can be in a PC file, floppy disk, CD-Rom, etc. If the source code is lost, it can be virtually impossible with some systems to reconstruct source code from the compiled code. This disadvantage is also true of graphical programming where the input is always stored off-line and compiled. Modern systems can combine back-up utilities which are advanced enough to save all of the program contents to a number of storage options. Also, employees come and go on both sides, so keeping backups in a company safe-place is prudent In today's complex programming systems, a back-up copy typically ends up on a hard drive or CD-Rom. If the operator station hard drive fails how do you have a backup ready to load on another machine?

For a system of any significant size it is advisable to regularly make an additional copy of all the site-specific programs onto storage media and have

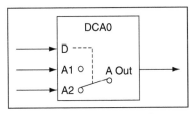

Figure 10-10 DCAO Gate

them stored at a remote site. It only takes one fire in the operator's room to discover what good insurance having the programs at a remote location would have been!

Application-specific programming is much simpler from the users's perspective because the manufacturer already writes the actual control logic. The resultant graphic is much easier to look at than just words on a page. It is usually presented in a standard format for each type of application specific controller. The prewritten programs are generally burned into a chip in the microprocessor which can be programmed but the program cannot be changed, PROM, so the programmer may not change them. The user interface usually takes the form of question-and-answer (Q&A) or fill-in-the-blank templates. For instance, for a standard single-zone unit control, the user would tell the control system: whether the system has cooling or heating capability, or both; whether two-position or modulating control is used; whether the system has an economizer and what type of high limit lockout switch is used; what set points are to be used; etc. The Q&A interface can also be graphical; the user would bring up a pre-developed schematic of the system on the PC, click on (using the computer's mouse) the devices that are actually present in the application, and then enter configuration and set point details through pull-down menus.

Both general purpose and application-specific programming have advantages and disadvantages. General-purpose programs are custom-written to achieve almost any desired control sequence, even the most unusual or uncommon. This provides a great deal of power and flexibility, but it requires more knowledge on the part of the programmer and increases programming time. It also increases the time required to commission the system because custom software is more likely to contain errors, called bugs.

Application-specific software, on the other hand, requires much less time to program and, because the programs are debugged by the manufacturer, usually requires much less time in the field to commission. However, preprogrammed software is limited to the control logic burned into the chip. Therefore, if a given application does not exactly match the control concept programmed into the controller, some compromises must be made and the system may not perform as well as if a custom program was developed using a general purpose programming language.

To avoid this compromise, sophisticated users can sometimes "fake-out" the preprogrammed software into operating as desired by adjusting set points and otherwise using programs in a manner different than originally intended by the manufacturer (sometimes referred as "work-arounds" or "alternative uses"). But this defeats the simplicity of application-specific programming, and is not recommended. This can be very difficult to document and support. Several manufacturers have the flexibility of allowing both types of programming; prewritten prewritten programs are provided for common applications, but users can also write their own programs using a general-purpose language or "user-defined programming" to connect into the application program. This methodology of "start with this known reliable base and add to it if you really need to" is becoming the standard approach in the HVAC controls industry. You will see it again when we discuss communication protocols for networking in the next chapter.

10.5 Specific Programming System Features and Parameters

Internal to the controllers are software routines that perform specific features. Usually they are menu-driven for ease of use. Scheduling options are shown in *Figure 10-11*.

Start–stop loops and scheduling is made to points that need to be started and stopped such as chillers, pumps, AHUs, and exhaust fans. A time schedule is set up and the particular equipment BO (binary out) point is assigned to that start–stop loop. In addition, daylight savings time is accounted for, and holidays/special event days can be programmed for running. Internally the start–stop points are staggered to start and stop to achieve timing balance such that motor loads are not all started at the same time avoiding kW demand challenges. Anti-cycling timers are also available, so that the equipment does not start and stop too many times during a given time period. As you can see from *Figure 10-11*, a single schedule can be assigned to control an assortment of equipment and multitude of loops.

Clock synchronization and network time are established and maintained. Each controller that uses time as an input should have an internal clock to keep track of time. These clocks in control panels and workstations are automatically synchronized daily from an operator-designated device in the system. Time could be synchronized on a regular basis, typically every day, or upon the event of a time change or power loss. The system shall automatically adjust for daylight savings and standard time.

Software interlocks are performed a number of ways using the controller software. Typically, a start–start loop is created, so the equipment to be interlocked has a BO output point (the BO typically starts a fan motor or system) that is assigned to that start–stop loop, and it is programmed to go "on and off" with that start–stop loop configuration. The BI shown as a bypass can cause this BO to be turned on when it is scheduled off (unoccupied), and the BI disable can cause the BO to go off when it was scheduled to be on. A number of time delays and logic can be programmed to modify its speed of response.

Bypass override is the software feature that overrides the normal time scheduled unoccupied time that a controller point is scheduled with and allows operation of that point for a specified amount of time, depending on the amount of minutes programmed (*Figure 10-12*). It is very useful for operators that use the time schedule programs for their systems, but sometimes have the need for the system to run during unoccupied times for cleaning, overtime work, etc. Tenant override is a variation of this feature, and it is used a lot in office buildings so that the tenants can come into the building during unoccupied times, activate the control system for the time they are there. The software also keeps track of who overrode the system and for how long, so that costs can be assessed for that tenant.

- *Pseudo, or Virtual, Points.* These are software points, loop points, or artificial points for holding information. In the tenant override situation, the system will have a pseudo point, or set of points, to store the accumulated usage time. Pseudo points may be single value or a defined table of values. For trending the pseudo point could be a table defined as having a column for the time of sample and a column for the sample value. Each time the

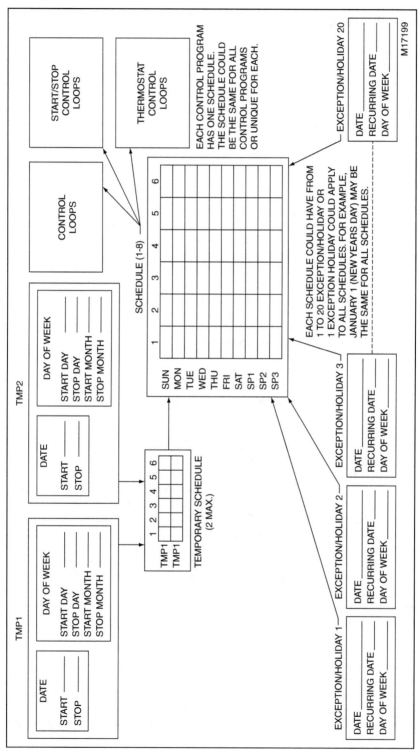

Figure 10-11   Schedule Components

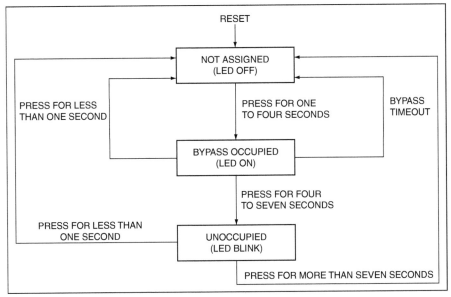

Figure 10-12   Bypass Logic

trend took a sample it would be added, along with the time, to the table until the allocated table space is full.

Figure 10-13 shows uses and positioning of controller I/O points being used in a software control loop depiction.

Lead–lag programming allows the operation of equipment to change its "first to start" routine and makes all of the programmed equipment operate on a more even time basis. When used in conjunction with other control

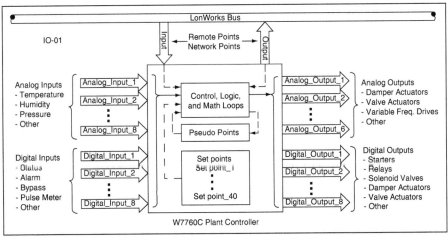

Figure 10-13   Controller Points

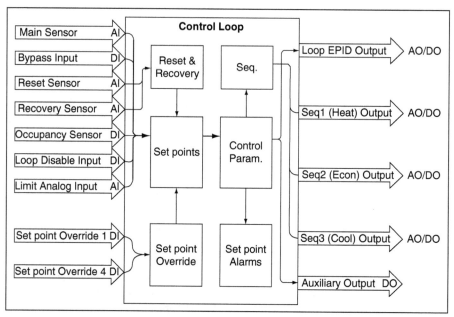

Figure 10-14   Control Loop

functions, the lead device comes on first, but if the second or third device is needed, they can be brought on automatically.

Control loops are used to control small control systems, *Figure 10-14*. For instance, a control loop for discharge air control would use the discharge air temperature as the main sensor input, and control valve signal as its Loop EPID output. The software that you program into the control loop looks at the input data, and then makes adjustments to the output value in order to maintain its set point. The parameters that are programmed by the user into the control loop are set points, override times, PID values, and action direct or reverse, set point resets, inputs and outputs, interlocks.

Staging of heating and cooling outputs is very commonplace. Typically, the stages of cooling and heating devices are set up so orderly and even temperature control can be accomplished.

Sequencing of cooling, heating, and economizer functions are programmed into this feature. When used, it allows for a smooth transition to exist between these modes and does not allow one mode to run while the other mode is working. For example, it prevents cooling and heating to run at the same time. Remember from other chapters; sometimes during the de-humidification process, cooling and heating processes occur simultaneously. There is a lot written about PID loops in previous chapters, but by using DDC systems, this programming is primary in its functionality. *Figure 10-15* demonstrate the actions and reactions of the PI programming options frequently used in DDC systems. The "D" of PID is purposely left out as it is rarely used in the control of HVAC processes.

Using a PI routine, *Figure 10-15*, the proportional error is smoothed out over time, and the integral time helps get closer to set point as time goes by.

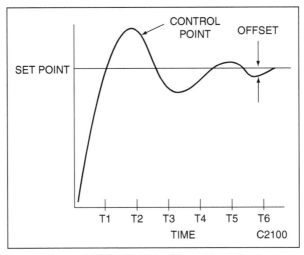

Figure 10-15 PI-Control Loop Time Graph

Math functions are set up in the software to take analog values for real-time data points such as AIs, and put them into a utility that can add, subtract, divide, multiply, average, minimum, maximum, high select, low select, etc., as indicated in *Figure 10-16*.

The output of this math function can then be used in a control, start–stop, or a logic loop to provide an output result.

Set points are programmed into the controller and assigned to the control, start–stop and logic loops. They can be hard numbers or can be a function of a reset routine or math function as discussed above. The software can be programmed with set point limits that will only allow the set point to be between the values specified; the best example is a thermostat on the wall with a set point dial that says 55–85°F; that dial can be moved to any value, but when you use set point limits, the controller will only see the limits. Alarms can be set up so that if the user moves the set point of a given device too high or too low, an alarm will be generated. Set point overrides can be set up so that a trigger mechanism tied to a digital input BI, can transfer that different set point value to be used in the algorithm.

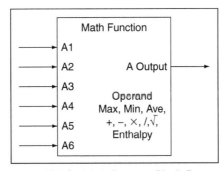

Figure 10-16 Math Function Block Diagram

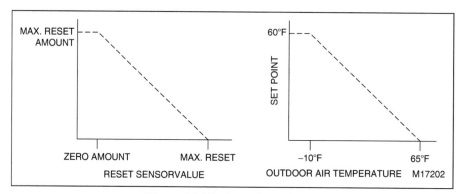

Figure 10-17   Reset Ramp Graph

Resets to the set point can come from many places. The reset function allows the set point to be changed in relation to another analog value that is occurring from within a controller: at one reset amount number, the reset is a zero amount, and then as the reset amount changes, the reset goes towards its maximum value (*Figure 10-17*). As a practical example, outside air temperature reset is common, where the discharge air temperature set point of 55°F is "reset"'s to 75°F when the outside air temperature varies from 75°F to 55°F.

Some control systems use what is called a form of "intelligent" setup and setbacks, whereby they have a "stand-by" period of time when the set point goes to its midrange value. The ramping function allows the program to move at a programmed speed when it changes from one set point to another. A typical example is that when a time schedule changes the mode from unoccupied to occupied, the set points programmed are sometimes different. The ramping function gradually changes the effective set point that the controller is looking at for a time interval. This allows for a smooth transition from one set point to another. A cooling reset routine, *Figure 10-18,* for VAV terminals looks at the common discharge air temperature coming down the duct and resets the operation of the DDC VAV terminals. This is also very commonly used in zoning systems, where by the thermostat action needs to be changed from direct acting for a cooling operation, to reverse acting for heating operation, automatically based on the change in the duct supply air temperature.

A variation of this comes with popular routines called recovery, morning warm-up\cool-down and evening setbacks or night purge, start and stop time optimization. An internal routine looks at the time each time the mode changes on how long the process takes to get to set point. *Figure 10-19* graphically depicts an optimal start routine.

Assume that it takes 15 min after the system is turned on in order for the process to make its set point. This recovery routine looks at that every time and establishes a historical time period that it should turn on "early", or "optimally" in order to make the set point at the specified scheduled occupancy time.

It continues to do this daily so this time period that it turns on early can change with the seasons. Similarly, the evening shut-down routine looks at

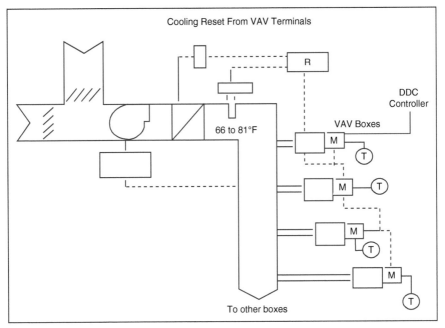

Figure 10-18 Reset from a Common Duct Temperature

how long it takes after the control system is shut off in order to reach it unoccupied set point. This time value can then be used perpetually to turn the systems "off early" to achieve its goal of reaching its unoccupied value at the specified scheduled time. This evening shut-down routine is not commonly used, but the morning one is. Some manufacturers have their own recovery routines; however they all do essentially the same thing.

Minimum and maximum run times and interstage delays work together to limit the time period in-between functions of the controller. The minimum and maximum time limits supervise the operation of a particular output or start-stop loop, and for the minimum run time it leaves it on for the min-time specified even if someone, or another program tries to turn it off earlier. The same holds true for the maximum, once the output is started, it will only run for that amount of time, and it will turn off until it is called for again. A common example is that of a pump or any motor load; once the pump is started, it should not be short-cycled so the minimum runtime is set up. A variation of this is the interstage delays and timers that are used as anti-cycle timers. In a staged application, the control system stages outputs on and off; the interstage timer specifies the time limits that the controller will bring on and off those stages. These are commonly used in DX and refrigeration systems where short cycling can be damaging to the system and can cause nuisance alarms and manual reset shut-downs.

Post and pre-delays or also known as On and Off delays are programmed into controllers to wait the specified time before or after a command is given before it allows that action to execute. For example, let's say that you are programming a chiller plant and its pump interlock operation. You want the

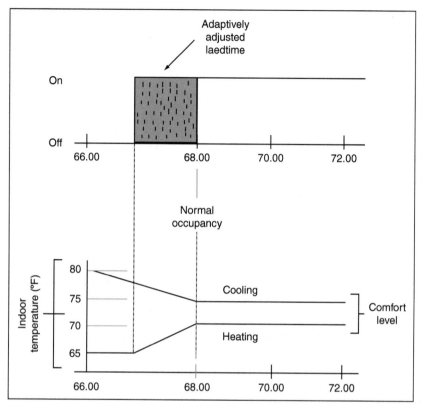

Figure 10-19   Optimum Start

pump to run for 5 min after the chiller shuts down. Then you would use a post delay or off-delay timer that would keep the pumps operating for an additional 5 min.

Demand Limiting is used in conjunction with kW/kWh monitoring of the power input to the building or facility. This same concept can be applied to gas or any other energy source. Site demand can be controlled by monitoring its sources, such as distribution, and utilities. The graph in *Figure 10-20* depicts the level of demand, and shows the "demand limit" point where the demand exceeds it set point.

The control program looks at the kW demand value and calculates it kWh over the time period. Demand kW set points can be inputted to the program, and specified control points that can be turned off in sequence or set points that can be raised or lowered. During the specified time periods programmed in the DLC (demand limit control) routine, this control system utility will turn off the specified loads and/or alter the set points as specified in order to limit power consumption to the specified kW demand set point. As the kW demand returns below set point the program will return to its normal conditions. This is a really useful cost-saving feature but is generally not used because of the facility typically cannot function without its HVAC&R. Software programs can track and record energy usage on a DDC system.

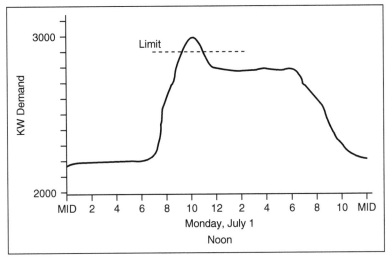

Figure 10-20   Power Demand Graph

Outside air temperature, relative humidity and CO_2 conditions are monitored and can globally be assigned to control functions and make adjustments that control system operation. Large chiller or boiler operations can be adjusted for its seasonal impacts. Also, outside air ventilation can be adjusted.

Other building systems can provide input signals to the DDC system for information and action. The fire alarm system can send the DDC system an input that it is in alarm, and the DDC system can shut off the air handling systems in concert with the fire alarm system. Most DDC systems are not listed/approved for the primary shut-down of the air handling equipment (because fire alarm and life safety NFPA standards/codes do not allow for it), but can certainly be used as a secondary source of shut-down. Security systems can communicate with the DDC systems and let it know when people are in the building; this feature is sometimes used to start or stop the HVAC&R systems on actual occupancy sensing and after-hours override situations. Lighting control systems can communicate with the DDC system and talk to each other about occupancy strategies and optimization.

Trend logging, *Figure 10-21*, is considered by many to be one of the most important features of DDC systems. This term refers to the ability of the system to store the values of input and output points on either a time-of-day basis (such as every 30 s) or a change-of-state basis (such as every time a binary point turns from on to off or vice versa, or every time an analog point changes value by more than a preset range). With many early systems, trend data were stored on the disks of the operator workstation computer. However, most systems now store trend data in the protected DDC panel memory or in a dedicated trend storage device connected to the communications network.

Trend data can be used to diagnose program bugs and to optimize the control system in general. With standard pneumatic or electronic controls, problems are often difficult to diagnose unless someone actually observes the fault in real-time, which is often a matter of luck. Trending removes this

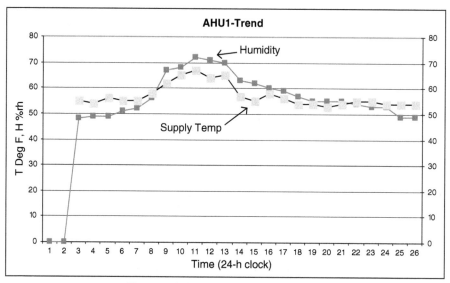

Figure 10-21   Trend Reports Example

element of luck. Some systems are able to present trend data graphically, often with several variables displayed simultaneously, or the data can be exported to other programs or spreadsheets for better visualization. Typical reports from a majority of DDC systems are named all points, Runtime, Alarms, Locked out points, Energy, Device status, Bypassed points, Trends, and Schedules.

Having discussed these reports, where do they show and what is the operator terminal like?

10.6 Operator Terminal

Operator terminals provide the human operator with access to the DDC computing processes. For setup and access to a single application controller it can be a small LCD display accessed by a set of buttons on the device, a smart thermostat being a simple example. Moving up with application specific controllers the terminal may be the controller manufacturer's hand-held LCD display unit which has character lines long enough for brief text questions, input and output. These terminals are adequate for application specific units where the tasks are setting up the unit, occasional set point modifications and checks.

When we progress to terminals for the fully programmable controller we move to laptop and desktop computers with full screens. For the maintenance staff, plugging a laptop into the communications port on the controller in the mechanical room and doing checks and making adjustments can make problems far easier to resolve. For the operator managing the facility a desk top computer is the norm. This desktop will normally be where all the programming is entered, downloaded, altered and stored. It is also the device which provides the operator with a view of the system.

Typically, systems offer two types of system view, text and graphic. The text mode enables the more system competent operator to check on performance by listing or finding points based on particular characteristics. "Which supply fan points are in manual control?" allows a quick check to be made that no supply fans have inadvertently been left in manual control. Almost 30 years ago when DDC was starting this was the only access for the operator and they had to learn the commands and recognize the terse text responses.

Life is easier now as most systems also have graphical interface. Each mechanical system is shown on a graphic which may well be a 3-D image, see *Figure 10-22*. This image is typically in color and likely automated in two ways. First, color change is used to indicate plant status. For example, a red fan is off and a green fan is running. Fan blades may be shown as rotating when the fan is running and stationary when off. Second, live data on operation may be shown beside the unit such as duct pressure beside the duct pressure sensor, supply temperature and humidity in the supply duct, *Figure 10-23*. This real-time 'live' graphic can make it very quick and easy for the operator to see what is going on in a particular system.

Not only can the graphic provide live data to the operator but the operator can use the mouse to locate an object and make changes to the plant operation either with additional mouse clicks or keyboard entry. The screen has become a two-way window to the plant providing the operator with information and allowing the operator to make changes.

Most graphical software packages also allow for bitmap or jpeg formatted file-based photographs to be used. The photo image of the equipment or device is placed in the background of the graphic screen, and then the real-time data is shown in the foreground. On larger sites an aerial photograph of the site can be used as the initial graphic for locating buildings. The operator can use the mouse to click on the building and is then taken to a graphic of that building.

Graphic packages from different manufacturers can vary in complexity, availability of "canned" images, colors, speed, compatibility with common

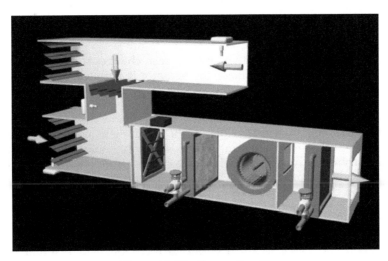

Figure 10-22 System Graphic

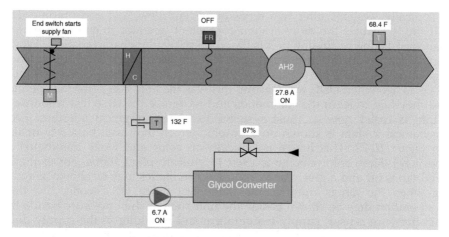

Figure 10-23   System Graphic with Plant Status Points

graphic generation tools such as Paint and Adobe, integration, colors, animation, and connectivity programs such as PCAnywhere, Laplink, and VNC (trademarked products).

Typically the beginning of the graphic displays starts with an Introduction, or Table of Contents menu, or Map of the area. Then it can show a "master" listing of mechanical–electrical systems, and sometimes-simplified floor plans of the building. For example, this might include each chilled water system, hot water system, chiller, boiler, pump, tower, valve, air handler, exhaust fan, lighting system, and all terminal equipment and this "master" page can have icons that will automatically transfer to that graphic for the equipment being monitored. Point information on the graphic displays should dynamically update. Ideally, the each graphic would show all input and output points for the system, and also show relevant calculated points such as set points, effective set points after resets, occupied status, applied signals such as voltage or milliamps, positions, ambient conditions, alarms, warnings, etc., but the resolution and size of screens means choices must be made as to what is actually included.

All this screen activity is lost when the screen changes so most systems also have a printer attached. This allows the operator to print either what is on the screen or text information from the system. A print of the screen with some hand written note on it can make it much easier for the operator to explain some odd behavior to the maintenance staff the following day.

In addition to the on-screen visuals of the plant, there are reports. The most important reports are the alarm reports notifying operating personnel of alarms, warnings, advisories, safety faults or abnormal operating conditions. Alarms are, ideally, prioritized in both transmission and display. A fire alarm should be on display in seconds but failure to maintain a room temperature can, in most cases, wait a minute or more. A challenge on larger systems is the number of alarms and warnings and how they are dealt with by the operator. Some method of suppressing display of lower importance alarms awaiting attention can make it much easier to focus on, and respond to, new important alarms.

Alarms can usually be accompanied with helpful/useful text. In one customer-considerate building with numerous tenants, the alarms had a tag number which identified the name and phone number of the particular tenant's contact person. This made it easy for the operator to phone if they thought the alarm could affect the tenants. In other cases the message could be diagnostically helpful. Earlier in the chapter we discuss how a current sensor and AI point on a constant volume fan motor can indicate overload and lack of load. When these alarms come up they can be associated with a message "fan over current" or "fan running, but low load" which gives the maintenance staff some assistance on evaluating the problem.

In addition to the graphics screens of the plant in operation and alarms the operator can set up and receive many other reports.

Trending is perhaps the greatest operator benefit of DDC. The ability to obtain a history of operation can be extremely valuable in detecting problems. A simple example: a library staff member phoned to complain that the library was hot one very cold winter morning. A quick check showed 72°F, the design temperature, so the complaint was logged but not acted on. The following morning exactly the same call and again the temperature is 72°F. This time a log is set up for the space temperature. *Figure 10-24* was the result showing a high overnight temperature. The staff member was an early bird and came into the hot space but delayed phoning until the office staff in the maintenance department started at 8.30 a.m. by which time the temperature had dropped to the required level. This was a building with perimeter heating controlled by outside temperature reset and an air system to provide ventilation and cooling. The perimeter radiation was set to high and overheated the space during the night.

For systems with fast recording capabilities, trending can be used to check on the performance of controls on startup to see how the controlled variable is controlled on startup and after a change in set point.

This ability to trend data from the system provides two big opportunities in fault detection and energy monitoring. Let us consider a simple fault detection

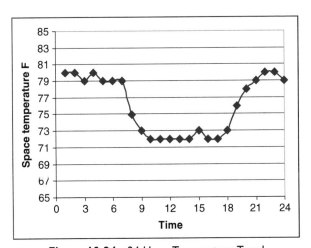

Figure 10-24 24-Hour Temperature Trend

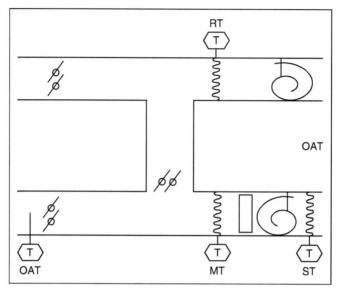

Figure 10-25   System Diagram with Temperature Sensors

situation in the air-handling unit shown in *Figure 10-25* which has temperature sensors for return air, RT, outside air, OAT, mixed air, MT, and supply
ST. Let us imagine that the system starts in the morning with the dampers
and coil valves remaining closed. It runs like that for 10 min before damper
control is enabled.

What would you expect in the way of temperature records after 8 min? RT,
MT the same, and ST a degree or two higher as it has the added fan heat
seems a reasonable expectation, whatever the outside temperature. If the system were set up to record this data each day, the temperatures should maintain a consistent pattern RT approximately equal to MT and ST a degree or
two higher.

Now suppose one hot muggy morning the data show MT and ST much
higher than RT. Either RT is reading low or the dampers are not properly
closed. Another morning RT and MT are still close but ST is lower than either
of them. Now either ST has failed or the cooling coil valve is open. The data
can be collected and the fault finding logic implemented to provide a degree
of self diagnostic fault finding within, in this example, 24 h. Now there is
the potential for the system to self diagnose and ask for the correct treatment!

In a similar but different manner some DDC systems also have a "live
demo" testing software available so that most of the DDC system can be seen
operating by a test database of values operating in simulation of real-time.

The other trending advantage is in energy monitoring. This can be used in
load scheduling as discussed earlier and also for tracking overall plant performance. Monitoring utility meters can provide useful information on how well
the plant is operating and whether usage is going up or down. A simple
example is the office building in a climate requiring heating in winter. The

heating requirement for outside air and fabric heat loss is approximately dependent on the difference between inside and outside temperature.

A reasonable approximation to the scale of this difference is to subtract the mean temperature for the day (highest plus lowest divided by 2) from a base temperature, commonly 68°F. This gives a number of degree days below the base for the day and is called the degree days for that day. Summing the degrees for all the days of a month provides a measure of how cold the month was. Now if one plots the monthly degree days against heating energy consumption (gas if heating is from gas) one would expect them to rise and fall together. An example for one year of data is shown in *Figure 10-26* with a trend line added. Now in a coming month the consumption can be plotted against the degree days and if the point is above the line it indicates a higher than expected consumption while a point below the line indicates a reduction compared to the history.

The degree days information can be calculated automatically on the system or obtained from the local weather office and consumption data accumulated from the meters for automatic display every month. One month's data are subject to errors but the trends can be very revealing and valuable management tool. Note that the data are consumption, not cost, so changes in rates do not affect the situation.

Most systems also allow trended data to be exported so that the data can be manipulated and displayed using standard text, spreadsheet, and database programs. A more complex use of trending will be discussed in the final chapter, along with other specification issues.

Operator terminals can be used to do almost anything with the system and few operators should be allowed freedom to do anything they like. Systems are thus provided with password security. Passwords are assigned levels and each level allows a range of actions. Alarms come up on the screen of the operating system that let the operator know of the alarm condition. If the operator password level is low, they can just see the alarm message and its instructions, but cannot cancel or 'acknowledge' it so that it goes away on the screen. Higher-level password operators can come in later and review the system actions and alarms so that a designated responsible person can be

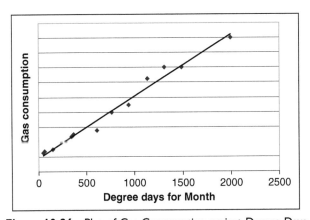

Figure 10-26 Plot of Gas Consumption against Degree Days

made aware of them. This feature is important so that alarms are dealt with and not ignored.

Most systems also record the operators as they log on and log off providing a history. This can be particularly valuable in a large operation where one wants to find out which particular staff was logged on last evening.

The control system software and front-end computer can also provide maintenance scheduling. It can be manufacturer provided or third-party add-on software. Programs can be provided that will print maintenance work orders and follow-up reports on a regular basis or after a certain number of equipment run-hours, alarm triggers or timed routines, thereby allowing improved maintenance and performance of the equipment. ASHRAE Standard 180 – Standard Practice for Inspection and Maintenance of HVAC Systems – is making several recommendations on using DDC systems for maintenance and operations of a typical building. Typical examples are filter changing, coil cleaning, and routine lubrication of motors at run time or time intervals are just a few examples of how proper care and maintenance of HVAC&R systems can improve energy efficiency and functionality.

Help screens and training resources are available from most DDC systems. The software documentation should be on paper but also available from the PC in software so that the operator can access it regularly. These help and training venues should also be updated periodically thru access to the manufacturers web page as automatic downloads.

This chapter has covered points, application programs, and operator terminals. The next chapter will cover how systems are networked together and the protocols that facilitate interoperability.

Bibliography

The bibliography for the three chapters on DDC controls is at the end of Chapter 12.

DDC Networks and Controls Protocols

Contents of Chapter 11

Study Objectives of Chapter 11

In the previous chapter, we assumed controllers and operator workstations could communicate with each other. In this chapter, we are going to consider the hardware needed to achieve this communication, and some of the software challenges and opportunities to make this communication work effectively. These challenges and opportunities are significantly influenced by rapid changes in technology, falling prices of microprocessor based technologies, and the common industry focus on lowest first cost often with little, or no, regard to cost in use.

This chapter introduces you to concepts. It does not aim to give you advice on choosing solutions. Rather, the aim is to empower you to talk to suppliers and manufacturers, to read their literature, and ask questions. Remember that this is an industry of jargon, so be prepared to ask what is meant by words, or numbers, you do not understand.

There are many good ways of designing a DDC system, although one may fit best in a particular situation.

After studying this chapter, you should be able to:

Understand how networks are structured and how information flow can be controlled

Be aware of the issues of interoperability

Have an understanding of the types of physical structure of networks
Know about the types of HVAC communication software and some of its
   similarities and differences

11.1 Interoperability

Interoperability is the ability of components and systems to work together. An
orchestral performance in a concert hall is an example of interoperability, with
people selling tickets, the orchestra practicing, the HVAC is set running, ush-
ers direct people to seats, the music is played ... This is a fairly simple, single-
focus example of a human interoperability situation. In this example, timing
is generally not just important but critical during the performance. Many mistakes can be dealt with by people helping out and the humans in the operation
are wonderfully flexible. They can make adjustments in-the-moment to
accommodate unexpected events.

In the computer world, the world of binary data flows through silicon and
along wires: there is virtually no flexibility. In the computer world, interoperability is an issue of exact matching. Computers are not as adaptable as
humans; they need things to be correct in every way.

For us to communicate using written language, we need to have at least a
rudimentary knowledge of the other person's language. Note that we may
think in one language and speak in another language. In the same way a controller can run any application software to run the control logic but it must
output data in the correct format, or language. The computer has a dictionary,
and words must be in the dictionary in order to have a meaning. You are
reading a text in US English, so people live in "apartments," but you may live
in the UK where the word for apartment is "flat." Even when the word is the
same, such as "color" it may be spelled (or spelt) differently – "colour." When
reading we can also deal with spelling errors. You can probably work out this
sentence: "This mcoputer is not intelligetn." but for the computer the only
words it knows in the sentence are: This ... is not ...

Not only must the words be right but also the method of transmission.
When we are speaking to each other we need not only a common language
but also to be loud enough and not speak too quickly, or slowly, in order to
be understood. This ability to understand each other depends on how alike
our dialect is and common speech speed. Computer transmissions are all
based on changes in an electric signal. Those signals must be transmitted
and received within limits for form, voltage, and frequency.

You know that the fact that two people speak English does not guarantee
that they will be able to understand each other immediately; even with practice they may have difficulty due to dialect, individual words, and colloquial-
isms. You therefore will not be surprised that interoperability between
computers is a daunting challenge of details.

The challenge occurs in two specific areas:

- the physical communication connection, called a network; and
- format and content of the messages sent and received.

Because networks are needed in every facet of the modern world, they
have been specified and standardized internationally. The definition of how

a network is to be physically constructed, how components are to be physically connected, how information is to be packaged, and how information packages are to be scheduled for sending, all form the network "protocol." The protocol defines the network in such detail that two manufacturers' equipment will, if both meet the protocol requirements, communicate. "Plug-and-play" is the current expression for this conformance – when it works well.

Unfortunately, the format and content of messages in HVAC are not so standardized and internationally agreed as the protocols for networks. Initially, as DDC started in the HVAC industry, manufacturers each developed their own unique, proprietary, application software to produce a DDC system for HVAC. These proprietary systems could not communicate with each other unless a "translator," called a gateway, was developed to translate the messages from one system to another.

These proprietary systems, now often called "legacy" systems, caused many challenges for users. Specifically, operators needed to interact separately with each system and to know how each one worked in order to do any maintenance. Theoretically, one could change a site with three legacy systems into a single virtual system for the operator, as shown in *Figure 11-1*. Each gateway acts as a translator, converting from a legacy language to, in this case, operatwor-terminal language. Developing each gateway is expensive and requires full details of the legacy proprietary language. This "solves" the problem for the operator but the maintenance staff still need to know how to maintain three systems or have maintenance contracts for each system.

Having paid a substantial sum for these gateways, let us imagine that legacy A manufacturer produces a new revision to their software, with some very attractive features. Unfortunately, this means redesigning the gateway to deal with the software revision, or not implementing the revision. These challenges of gateways and software revisions made many users push for more standardization in information transfer.

The result is a variety of protocols of which two, BACnet and LonWorks, are the major players. This chapter will be covering an introduction to some networking concepts, network types commonly used in HVAC, and communication standards for HVAC in terms of BACnet and LonWorks.

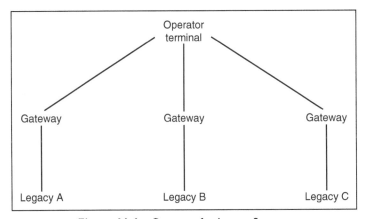

Figure 11-1   Gateways for Legacy Systems

11.2 System Hardware Architecture

When DDC systems started to enter the commercial and institutional HVAC control, 30 years ago, computers were very expensive and telephone cable was relatively inexpensive. These first systems typically had a few general purpose computers programmed to be controllers, and long multi-core cables to input and output devices. The communication requirements between computers and the operator workstation were not great, so a simple, dedicated, network worked well. Each controller computer was hard wired to numerous I/O points, as generally indicated in *Figure 11-2*.

In the following years, the relative cost of controllers dropped substantially and the cost of installing cable rose. The reduction in controller cost was partly due to the fall in microprocessor cost but also because manufacturers were making the microprocessors specifically as HVAC controllers instead of programming general purpose computers to do the task.

This reduction in cabling to points is well illustrated in the VAV box controller. The VAV controller is now a microprocessor card mounted in the same housing as the damper operator. It has few inputs and outputs all to do with the one VAV box in its immediate area.

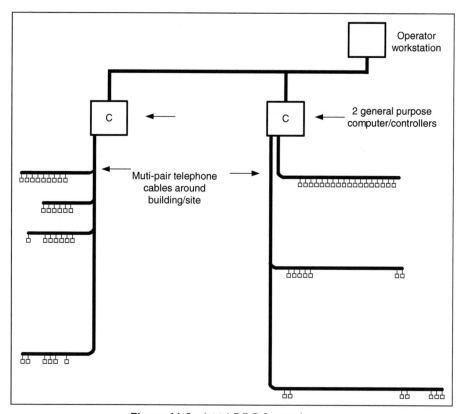

Figure 11-2   Initial DDC System Layout

The large increase in controller numbers and increased communication between controllers would make the traffic on the network so high that a really expensive network would be needed. To help control traffic, two devices were introduced: the bridge, and the router. A bridge simply separates two sections of a network and can selectively allow traffic through. For example, consider a system with numerous VAV boxes. *Figure 11-3* shows two branches of a network. The VAV box controllers can all talk to the master controller, C, for their segment, but only the master controller data are allowed to pass by the bridge, B. In this way, the traffic between sections can be minimized, and the same type of economical network used for more of the total system.

Many manufacturers now produce controllers that include the bridge function in the same enclosure. The controller is provided with a communications port to connect to the VAV controller network, and a separate port to connect to the network to other controllers and operator workstation.

A device called a router is a more advanced bridge, as it will also take messages from one network cable using one protocol, repackage the message and send it on along a second network with a different protocol. A router connects, controls, and converts the packaging as it connects two networks. The router allows smaller, less sophisticated, networks to be connected, as shown in *Figure 11-4*. Here groups of controllers are on their own network communicating as much as they wish between themselves. The router allows messages

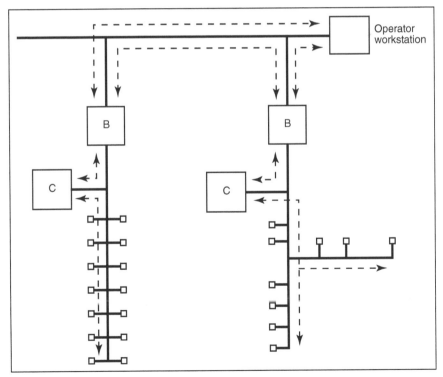

Figure 11-3 The Use of Bridges to Contain Network Traffic

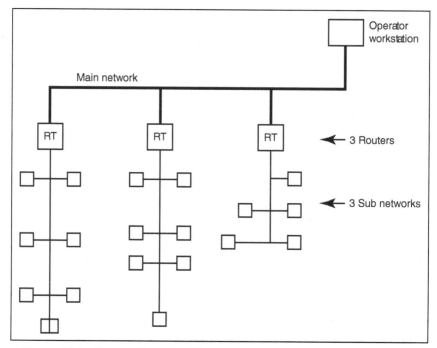

Figure 11-4   Routers Control and Facilitate Traffic Between Different Networks

from its sub-network to pass out to the main network only if they are needed elsewhere, and only allows messages relevant to the sub-network into the sub-network. This allows less costly sub-networks to be used for local connection and higher speed, and more costly networks to provide a communications backbone through the building or site.

As with bridges, manufacturers may package the router into a controller with two different communication ports. This will make one controller do double duty: as controller and router. The difference between *Figure 11-3* and *Figure 11-4* is that the network type is the same throughout in *Figure 11-3* using bridges but changes in *Figure 11-4* using routers.

The advantage of the layout shown in *Figure 11-4* is that the sub-network can be relatively low speed and inexpensive, while the main network is much faster and more costly to connect to and cable. The connection cost is higher as the transceiver that generates and receives signals is more sophisticated and has to run at a much higher speed. Typically, the sub-network will be twisted telephone cable, whereas the main network cable may be a more expensive, shielded cable.

The ways that messages are transmitted on the subnet network and the main network are often different. The router, therefore, not only decides which messages it is going to pass on but also has to package the message correctly. This repackaging is very similar to one having to use a special envelope to send a package if one is using a delivery agent such as FedEx or UPS instead off the general mail service provided by the National Postal Service.

Things get even more interesting when one needs systems to communicate over large distances. Typically, the telephone, or internet, will be used. For the telephone we use two half-routers, called modems, to send the signal. This is shown in *Figure 11-5*. Here the first modem (half-router) converts the network signal to an audio signal, which is transmitted by the phone system, which, at the other end, the modem listens to the incoming audio signal and converts it back to a network signal. The telephone call can be initiated by the controller and it provides a reliable method of raising alarms and doing simple operating checks. Transmission is very slow, so it can only send limited amounts of data.

Nowadays, there is the internet for long distance communication; the sending device converts the network signals to an Internet Protocol, IP, and signal for transmission over the internet. It can then be picked up anywhere in the world by an internet connection. The amount of information that can be transmitted over the internet is very high, but reliability in many places is not as high as the telephone.

The network in *Figure 11-6* has two systems by two manufacturers. Manufacturer A has an internet connection into one of their controllers, which could provide them with complete access to all of their part of the system. The operator terminal can communicate with all manufacturer B controllers, while manufacturer A controllers' gateway can translate and pass on the data. Manufacturer B can communicate over the internet with their system and with the data that is passed by the gateway. Telephone connections are a slow means of data communication, however they do have advantages of reliability and the ability to automatically dial out. Thus, a maintenance staff person can be phoned in a particular office, at home, or on a cell phone when particular

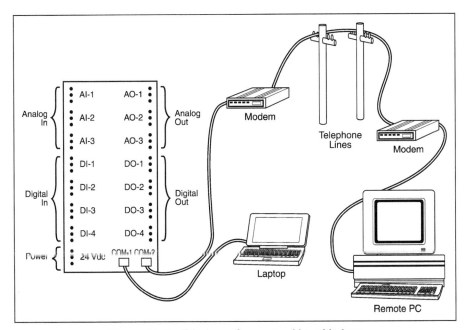

Figure 11-5   Telephone Connection Using Modems

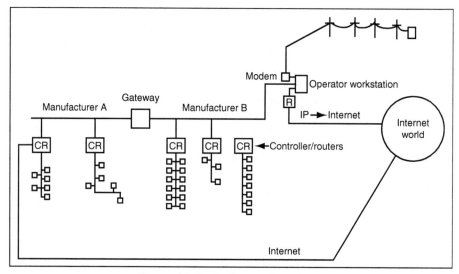

Figure 11-6   An Extended Network

alarms arise. Now with the internet capable cell phones, the DDC system can phone the right maintenance person and ask for help, and the maintenance person can log on through the internet and have access to the whole DDC system. The possibilities are endless, but the cost and complexity may not be affordable, warranted, or sustainable in a particular situation.

Another technology that is now available is radio. Local radio can be used instead of cable to provide network connections. A radio controller can send out a radio signal with the information encoded and a wireless receiver can detect that signal. One can therefore have controller-to-controller transmissions, and control-to-device transmissions. Controllers are typically powered by a 24-V supply (which can be a built-in transformer from the building power supply) but sensor devices may be 24 V or battery powered. When battery powered, the sensor device must be extremely low powered for reasonable battery life, so this is an emerging technology. The protocol used in the radio system is designed to link into standard building control network. The controller behaves, in effect, as a half-router, or modem, but instead of sending an audio signal along a telephone wire it is broadcasting a radio signal for another device to pick up and, with a half-router, convert the signal back. The difference is that the phone is point-to-point and wireless is often point-to-points.

Now that you know about the networking components, let us consider the system architecture, or topology, which refers to a diagram that shows the location of all of the control panels, computers, intelligent panels (where the microcomputer/microprocessors are), the physical location of I/O points, and how all of these devices communicate with one another. This topology can be very helpful in making decisions about controller size, and distribution and network traffic control.

Two major parts of the DDC system architecture are the hardware and the software.

In general, the hardware can be described as all of the control devices that make up the system – the PC, laptop, displays, controllers, sensors, relays, timers, transformers, etc. – the parts of the control system that you can touch and feel. They are typically depicted on an architecture diagram that outlines the hardware configurations in a broad-brush fashion.

Architecture diagrams can take on many forms (see *Figure 11-7*) Some show the layout of the hardware in a composite manner, and some, *Figure 11-8*, show how the hierarchy of the system lays out with the communications of the Building Management Systems (BMS).

Depending on the size of the DDC system, and how much integration into different manufacturers' controllers and protocols is provided, the architecture can be as simple as a one-line diagram or may be complex involving multiple levels of communication protocols and Operator Workstations. Also included in the architecture would be the ties to outside phone lines, intranet or internet.

The simplest DDC system is a stand-alone panel with input and output, I/O, points included as an integral part of the panel. The term *stand-alone* means no communication with other controllers is required for the system to work; the panel includes all the required components, such as microprocessor, memory and I/O boards. Optional features include a timeclock (usually required for on/off time scheduling) and power supply (required either within the control panel or as an external accessory). However, in practice, even the smallest DDC controller will accept more I/O points. The actual I/O point count varies widely among manufacturers.

Note that the degree of stand-alone ability can vary. As an example, consider a controller operating an air-handling unit with an economizer damper

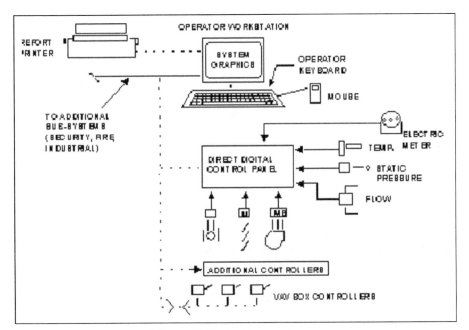

Figure 11-7   Architecture Layout

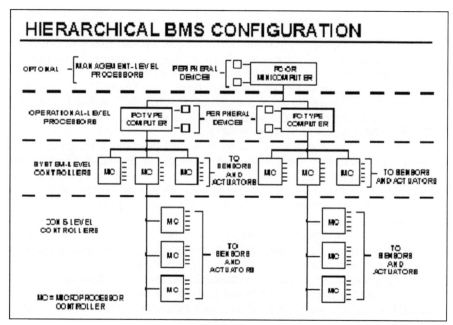

Figure 11-8   Hierarchy

system. The control uses outside air temperature supplied from another controller to switch from economizer operation to minimum outside air. In the pure sense of the term, the controller is not "stand-alone." However, if the software is written so that the economizer reverts to minimum outside air if the outdoor air temperature signal is not available, the system with continue to run. The possible energy savings of the economizer are lost but many would consider this to be a satisfactory "stand-alone" operation.

The enclosure that the controller fits into is important and should have adequate room to put in wire track or some way to make the wiring tidy and aesthetic. Also, it can protect its contents from external elements: mechanical damage, weather, dust, and pests. Depending on the use, these panels may be NEMA 1 (general purpose), NEMA 12 (dustproof), NEMA 4 (weatherproof), or NEMA 7 (explosion-proof).

Typically, intelligent panels have stand-alone capability with their own internal timeclock, so that if network communications are lost, each panel can continue to operate on its own. If one panel relied on information from another (for instance, if readings from an outdoor air temperature sensor were shared among panels) and if communications were lost or if the panel to which the common sensor is connected failed, then the other panels could continue to operate using the last valid data reading, or fall-back operation, until communications were restored.

Distributed control allows each controller to be monitored from a single point (for example, using a front-end PC), so that data can be shared among controllers.

Stand-alone DDC panels, shown in *Figure 11-9*, show their "peer-to-peer" network tied to a central front-end PC. This is a typical building application

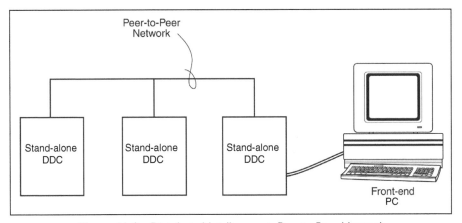

Figure 11-9   Distributed Intelligence + Peer-to-Peer Network

where the owners can easily monitor and operate their buildings independently of the controls vendor after initial installation and setup.

Some systems have two-tiered and multi-tiered communications architecture with this design: intelligent controllers are connected together by a
high-level, high-speed local area network, often using general-purpose LAN
protocols such as ARCNET and Ethernet. The front-end PC usually connects
to the control system on this LAN. The controllers then connect to and man-
age other lower level, lower speed LANs, often using a protocol proprietary
to the DDC manufacturer. To this second network, a combination of special-
purpose and multi-purpose controllers may be connected, each sharing information with the network controller, which can then pass the information on
to other systems on the high-level LAN. The concept of large "enterprise
wide" facility operations for multi-site locations is driving new and complex
requirements for BAS systems.

Sample Controllers

Controllers come in many configurations. A general all-purpose type controller that has a large number of outputs and inputs in a single controller panel is
shown in *Figure 11-10*. The physical unit is shown in *Figure 11-11*.

Some application-specific controllers have small and compact point densities
with small amount of outputs and inputs, usually just enough to perform their
application. An example of a VAV box controller is shown in *Figure 11-12*. Note
that the thermostat has a network access port, in this case called E-bus network
access. This allows maintenance staff to plug a communication unit into the
thermostat to monitor, change, and control the controller. This allows the maintenance staff communication access without having to, typically, get above the
ceiling to the VAV box controller. Note also that the controller housing also contains the damper motor minimizing installation labor.

Typically, the hardware is laid out so the wiring is simplified for the field
installer. Notice the wiring information given at the controller device level
itself. This is very helpful to the installer and maintenance technician. *Figure 11-13* shows a room thermostat, a true application-specific controller.

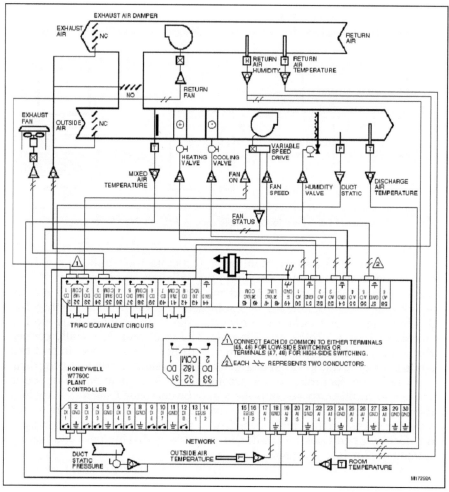

Figure 11-10 Typical General-Purpose Controller Diagram

It is designed to accept signals from several devices which may not be used including a motion sensor and humidity sensor. The device is designed to deal with the most likely requirements and is produced in quantity. The added hardware cost is very small compared with buying a general purpose controller and programming it for the specific situation.

11.3 Network Standards

Let us spend a moment revisiting the meaning of the word protocol. In these HVAC control systems there are two types of protocol. The first is the protocol for the HVAC information. Each controller needs to understand the meaning of the data being sent and to understand how to request data from other

Figure 11-11   Typical Controller Hardware

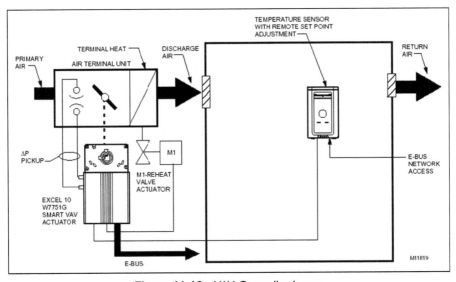

Figure 11-12   VAV Controller Layout

controllers. The second type of protocol is the network protocol. This is the standard rules for the packaging and unpackaging of the information to be sent and how it is physically transmitted.

Different networks have different costs and performances. In general, higher speeds and higher capacity comes at higher cost and need for more competent installers. Performance factors that vary include speed, capacity, and reliability in delivery within a defined time, number of possible nodes

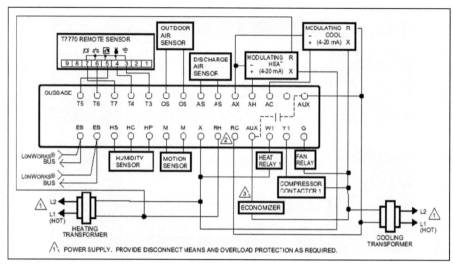

Figure 11-13   Room Thermostat Diagram

on a segment, and number of possible nodes in the system. These factors combine to provide networks with different performance characteristics. The cabling and required transceivers to put the messages into and receive messages from the network also vary in cost.

Network performance drops seriously as the network gets close to full capacity. It is important to specify that the network designer leaves spare node and traffic capacity. This will allow for the addition of nodes needed to make the system work as intended and allow for minor additions during the life of the system.

The main network standards used in HVAC are:

Ethernet (IEEE 802.3)

Ethernet is commonly used in local area networks, LANs, in organizations for general information systems under control of the IT department. There are three main versions, providing varying speeds and allowable distances: "standard" uses thick coax; "thin" Ethernet using RG-58 coax 100 Mbps 200 m; and unshielded "twisted pairs" 10 Mbps 100 m. The twisted pair version is limited to 100 nodes per segment. Segments can be joined with repeater. A repeater boosts the signal power and passes messages both ways but have no intelligence to restrict traffic.

If the site is to use Ethernet for general purposes, why not use it for the HVAC system? One reason is that the IT department may not want it used for your system! Why should they let you add a whole set of pieces of equipment to their network, which will produce somewhat unknown traffic and possibly produce serious security breaches? If using the site, Ethernet network is to be considered – the IT department must be involved from the very beginning. Even then, there will be difficulties as the terminology for devices is not consistent between the IT industry and HVAC industry: another language problem to overcome.

One particular issue to deal with is the uptime on the site network. Many IT-run systems are down for a time, typically at night or the weekend when changes, upgrades, and maintenance are done. Will all, or part, of your system be able to run effectively without communication for this period? If not, you will need a separate network. In many multi-building situations, most of the network can be "offline" for a few hours in the middle of the night, with just the powerhouse still connected by a separate network cable to the operator terminal. Note that any storage of trend data (storage of information about one or more points on a regular time basis) must be done locally during this network down time so controllers must have adequate memory capacity.

ARCNET

ARCNET is less expensive than Ethernet, running at 19 kps up to 10 Mbps on a variety of cable types. It is a simpler system, using smaller address and is limited to 255 direct addresses. An advantage, not typically important in HVAC, is that the time to send a message can be calculated. ARCNET is often used as the intermediate level in HVAC networks, Ethernet between buildings and to the operator terminal, and then ARCNET as the building network between major controllers.

RS-485

A very popular two-wire, plus ground, network for the lower level HVAC networks. RS-485, also the BACnet MS/TP protocol, provides for up to 127 master nodes and 254 nodes in total. This theoretical limit is not used, as communication traffic is too high for satisfactory operation. A limit of 40 nodes is the typical limit. The old limiting speed of 9.6 kbps has now been extended to 76.8 kbps. Master nodes can communicate with each other but slaves only answer when spoken to. The master node may then have an ARCNET communications port for the building network.

Wireless

This is a new technology, which we can expect to expand greatly in the future. Wireless communication has two separate features. The first, and as yet most implemented, is the use of the wireless signal to replace network cable. Each node communicates with one or more nodes using wireless signals. The second use is to enable devices to be independent of any wired power through the use of batteries. These two features of wireless are illustrated in *Figure 11-14*, where the battery powered thermostats, T, communicate with the building power controllers, C, which communicate between themselves and the operator workstation, OW.

The independent, battery powered devices, must be designed for very low power use. With current technology, this restricts their use to devices which receive and occasionally broadcast but are not involved in passing on messages. For absolute minimum power drain, receive-only devices may be used. The receive, and occasional broadcast, devices are often called limited function (LF) devices. A common independent device is a temperature sensor which can be placed anywhere within radio contact of at least one other node, and run independently for many years. In the minimum power format, it receives and responds to limited interrogation. This enables the designer to have the device

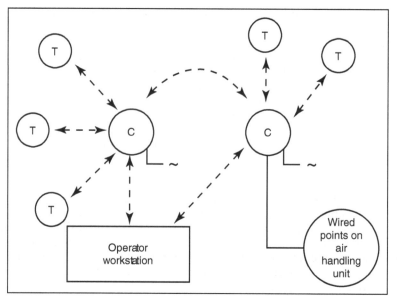

Figure 11-14   Wireless Network with Self- and Building-Powered Devices

in very, very low power use, most of the time extending battery life. Battery life with a lithium ion battery can easily be 5 years and may reach 10 years.

As circuits with lower power demand become available, this option will increase in popularity. Note that this technology will likely decrease first cost as wiring can easily be over 50% of the cost of an installed sensor. The cost of battery replacement includes the cost of the battery, finding the device, and replacing the battery. This maintenance cost needs to be factored into system support costs.

The full function, FF, wireless node is provided with building power and can pass on information as well as send and receive data. The full function nodes can operate in a linear arrangement, topology, where signals have a single defined route from node to node to node (*Figure 11-15a*). Becoming more popular is mesh topology, where full function nodes are arranged to be within

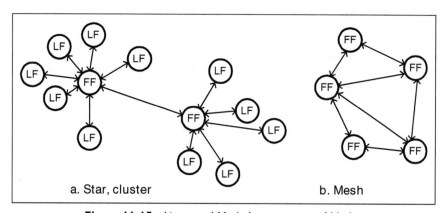

a. Star, cluster   b. Mesh

Figure 11-15   Linear and Mesh Arrangements of Nodes

range of at least two other nodes and messages can be passed by any route (*Figure 11-15b*). This makes for a robust wireless system not affected by interruptions in particular node to node paths as other paths will maintain communication routes.

How robust is robust? It all depends. Wireless signals are significantly affected by distance and materials – particularly metals. The signal attenuates (reduces) over a distance, and materials such as concrete attenuate the signal level by absorption. Metal acts as a reflector and the greater the area of metal, the greater the wireless signal is affected. Thus, a bus garage is likely to require nodes to be much closer together than a country hotel constructed of wood, studs and wall board. This attenuation issue makes the layout a try-and-see exercise. An experienced company can estimate the node spacing required. On installation, signal strength tests must be carried out to ensure adequate signal strength to, and from, each node to other nodes. If necessary, repeaters may have to be added or nodes moved. A repeater is a device that receives and rebroadcasts the message: effectively a signal power booster.

Wireless is very different from hard-wired connections. Interference can come from many sources, and an experienced supplier should be used for any significant installation. There is the building and its initial contents that act as permanent attenuators. In addition, there are the things that move – furniture, people, and new partitions – to allow for. Finally, there is the interference from other equipment such as lights, variable speed drives, elevators, and other radio equipment. The process is thus different from specifying a wire from A to B, installing it, and using it. With the wireless communication, an initial layout must be tested and often adjusted to achieve full operation. This adjustment will often be done after signal strength measurements have been taken on site to establish the wireless signal strengths.

There are three radio frequencies generally used:

- **2.4 GHz** frequency using the IEEE 802.15.4 standard. The system may be proprietary or, potentially, conform to the developing ZigBee™ Alliance standard (www.zigbee.com). This frequency band requires low power but is limited in data transmission speed. Indoor range in the normal commercial and institutional building for mesh arrangements is up to 150 ft while nonmesh communication may reach 300 ft.
- **900 MHz**. This frequency band has greater range than the 2.4 GHz band but can only be used in North America and Australia, which makes manufacturers' development costs more difficult to recover.
- **800 MHz** is used outside North America and Australia, and with the same greater range but limited geographic area of use.

11.4 BACnet

BACnet is a protocol dealing with the HVAC information flow around networks. It is an internationally accepted and approved protocol covered by ISO Standard 16484, Part 5: 2003 (EN ISO 16484-5), *A Data Communication Protocol for Building Automation and Controls Networks Systems*. In the HVAC industry there are three types of information protocol, with BACnet being the third type in the set:

1. Proprietary protocols owned and used by individual manufacturers. They are generally incompatible with each other, and one needs a gateway to translate information passing from one system to another.
2. Standard protocols, which are generally available protocols used by many companies but not subject to any internationally recognized standard organizations. Modbus is an example of one of these *de facto* standards. It has become very widely used in electrical switchgear and boiler plants.
3. A standard protocol, where the protocol has been very clearly defined, is publicly available, and is recognized by standard authorities such as the American National Standard Institute (ANSI) and International Organization for Standardization (ISO).

ANSI/ASHRAE Standard 135-2004 BACnet a Data Communication Protocol for Building Automation and Control Networks, BACnet, is a set of rules about controller hardware and software, specifically relating to communication for HVAC, lighting, smart elevators, utility metering, and Physical Access Controls (PACs). Over time, this list will, no doubt, increase.

The Standard is a large and very technical document, nearing 1000 pages with addenda. For the person new to BACnet, the ASHRAE BACnet Committee maintains a website at www.bacnet.org which provides numerous sources of information in a less daunting format.

The Standard contains rules about how information about points is to be represented, how the information is to be formatted for transmission, and how transceivers for different networks are to work. The information about points includes some mandatory items, several optional items, and proprietary items. The rules are open-ended in that there is choice in which are used and how to add options. For BACnet controllers to interoperate they need to comply with compatible sets of rules. The standard includes mandatory rules in a fixed format for the particular device. The rules define some mandatory communication data for the device, some optional communication data, and the freedom to produce proprietary communication data.

For example, a smart sensor must be able to identify itself (defined mandatory), it may have adjustable maximum and minimum end points for user adjustment (defined optional), and it could have the proprietary ability to report the highest temperature the user had requested (proprietary). For interoperability between this device and another device the optional properties can be requested but for proprietary properties both devices have to know the content and format of the proprietary information. If a BACnet system with proprietary content has been installed when it comes time to extend the system other manufacturers BACnet devices will not be able to access this additional functionality without obtaining the propriety details from the original manufacturer.

BACnet deals with the transfer of data between devices. It does not deal with the application program that runs the control loops, does stop/start, and all the other logic. It is similar to two people talking English to each other. One may be a native French speaker thinking in French and the other can be thinking in Spanish. Their "application programs" are different but they are converting the information into English words to exchange.

What is a PICS?

PICS stands for "protocol implementation conformance statement." It is basically a BACnet spec sheet containing a list of a device's BACnet capabilities. Every BACnet device is required to have one. It contains a general product description; details of a product's BACnet capabilities; which LAN options are available; and a few other items relating to character sets and special functionality. A PICS is the place to start to see what a device's capabilities are. Conversely, a specifier could draft a PICS as a way of conveying what BACnet capabilities are desired for a particular job.

- Basic information identifying the vendor and describing the BACnet device.
- The BACnet Interoperability Building Blocks (BIBBs) supported by the device (both the required BIBBs and any additional BIBBs supported).
- Which of the six standardized BACnet device profile does it conform to (note that additional profiles are under review).
- All nonstandard application services that are supported with details.
- A list of all standard and proprietary object types that are supported with details.
- Network options supported.

BACnet International (BI) was formed to, among other things, provide testing facilities so that manufacturers could have their device tested and approved as conforming to specific device profiles. The BACnet Testing Lab, operated by BACnet International, tests BACnet devices for conformance and interoperability. It lists devices that pass its testing, and only those devices are permitted to bear the BTL Mark. A sister lab run by the BACnet, Interest Group Europe, is operating in Europe. More information is available at www.bacnetinternational.org/btl/btl.php

This provides designers and users with a basic conformance which they can use to facilitate interoperability of BACnet devices from different sources. In addition, testing tools are being developed to enable automated testing of BACnet devices.

BACnet can be considered on four levels:

1. **Devices** – controllers, what is in them.
2. **Objects** – the constant and changing information in the real and virtual points in the device.
3. **Services** – definitions of how information can be requested and how information can be sent as needed or to answer a request
4. **Networks** – communicating with other devices.

I. Devices – Groups of Objects to Manage Activity

Devices are groups of points, called objects, plus instructions on how to ask for and receive information. A device might issue a request to obtain the current outside air temperature from a specific AI (analogue input) on its physical device or another physical device. The abilities of a device to request information from another device or to send information to another device are called BACnet interoperability building blocks, BIBBs. The simplest, the smart sensor has only two BIBBs. It must be able to report the values for any of its BACnet objects and must be able to accept an instruction to change the value of any property.

There are six standard BACnet devices, listed in decreasing order of communication ability:

BACnet Operator Workstation (B-OWS)
BACnet Building Controller (B-BC)
BACnet Advanced Application Controller (B-AAC)
BACnet Application Specific Controller (B-ASC)
BACnet Smart Actuator (B-SA)
BACnet Smart Sensor (B-SS).

For each device there is a set of required BACnet interoperability building blocks, BIBBs, functions that they must be able to perform which are listed in the Standard. A manufacturer is free to include additional BIBBs in a device.

2. Objects – To Represent Information

An object is a physical or virtual point with defined properties. It could be a temperature input, physical AI. This AI is defined as an analog input with properties: name, present value, status, high limit, low limit. It could also be a virtual object with a single value or matrix of values. Single value examples are point enabled, accumulated running hours, and set point temperature. Virtual points containing a matrix of values could be a time schedule or trend values.

In the 2004 Standard, there are 28 defined objects of which the device object is the only mandatory object in a device. It must exist and know what it is even if it does nothing! The device object includes information (properties) about the device, manufacturer, version, etc. Objects include the real analogue, binary, and pulse points as well as a variety of virtual points including calendar, trend log, event, schedule, and life safety.

A manufacturer may use the defined objects and may also design its own objects as long as the designed object conforms to the rules. Note that, for interoperability, the definition of the manufacturer designed object must be known, and readable for any other device to access the object. In the tables of properties for objects each property is designated as W, R, or O where:

W indicates that the property is required to be present, readable, and writable using BACnet services.
R indicates that the property is required to be present and readable using BACnet services.
O indicates that the property is optional, (but if present must return data as specified in the Standard).

For a seemingly simple object, an AI, a sample properties table is shown in *Table 11-1*, which is an example of an analog input object that is used for mixed air temperature of an air handler. The object supports both change of value and intrinsic (means based on a built in algorithm, or rule, such as a high limit) reporting.

Table 11-1   Example Properties of the Analog Input Object Type (from Standard 135-2004)

Property: R	Object_Identifier =	(Analog Input, Instance 1)
Property: R	Object_Name =	"1AH1MAT"
Property: R	Object_Type =	ANALOG_INPUT
Property: R	Present_Value =	58.1
Property: O	Description =	"Mixed Air Temperature"
Property: O	Device_Type =	"1000 OHM RTD"
Property: R	Status_Flags =	{FALSE, FALSE, FALSE, FALSE}
Property: R	Event_State =	NORMAL
Property: O	Reliability =	NO_FAULT_DETECTED
Property: R	Out_Of_Service =	FALSE
Property: O	Update_Interval =	10
Property: R	Units =	DEGREES_FAHRENHEIT
Property: O	Min_Pres_Value =	-50.0
Property: O	Max_Pres_Value =	250.0
Property: O	Resolution =	0.1
Property: O	COV_Increment =	0.2
Property: O	Time_Delay =	10
Property: O	Notification_Class =	3
Property: O	High_Limit =	60.0
Property: O	Low_Limit =	55.0
Property: O	Deadband =	1.0
Property: O	Limit_Enable =	{TRUE, TRUE}
Property: O	Event_Enable =	{TRUE, FALSE, TRUE}
Property: O	Acked_Transitions =	{TRUE, TRUE, TRUE}
Property: O	Notify_Type =	EVENT
Property: O	Event_Time_Stamps =	((23-MAR-95,18:50:21.2), (*-*-*,*:*:*.*), (23-MAR-95,19:01:34.0))

The required properties for this AI are:

Identifier – a unique number address which allows the device to be
Name –
Type –
Status flags –
Event state –
Out of service –
Units –

As the properties of objects are defined, the system is backward compatible in that the basic list is maintained with only additional data types being added. Note that properties are often grouped. In the analogue input in *Table 11-1*, the properties from "Time_Delay" to the end are not defined if the AI does not generate alarms.

A group of objects make up a device and, to be useful, they must be able to communicate. The way that this is achieved is through services.

3. Services – Making and Responding to Requests

BACnet is a client-server protocol. It is based on a device being a client and sending out a request. This request may be for information or to instruct another device to do something or change something. A request might be "send me the current setting of the damper" while an instruction might be "store this as your schedule" or "set the occupancy flag to unoccupied." These messages are called "services."

The 35 services are defined-in-detail ways of asking for data and defined ways of receiving requests and responding to the request. They are grouped into:

- Alarm and event services
  Alarm generation, alarm acknowledgment, request for list of current alarms. Event services include reporting based on change of value, a pre-determined event (e.g. every hour, damper setting above 90%) or combination of events.
- File access services
  Transmission of file changes for sending out changes to software programs.
- Object access services
  Read, write and change properties of objects.
- Remote device management services
  Used to control remote devices, TURN OFF, TURN ON, synchronize time, and proprietary messages and instructions.
- Virtual terminal services
  This allows one BACnet device to interact with the application program in another device as if one had plugged a terminal into the other device.

The Standard defines each service as BACnet Interoperability Building Blocks (BIBBs) listed in Annex K of the Standard. The single value request is the ReadProperty BIBB, DS-RP. It comes in two versions. The A version, DS-RP-A, is the initiate version. The B version, DS-RP-B, is the execute, or respond, version. So, a controller, or client, would contain DS-RP-A so that it can ask a temperature sensor, or server, containing DS-RP-B for the current temperature. In effect, one device must know exactly how to ask the question and the other device must know exactly how to interpret the question.

The BIBB includes a message number, what object (point) value is requested, and what property is wanted such as "present value." This BIBB will be sent out and the reply should be either the BIBB ReadPropertyAck which includes the message number, object under consideration, and present value or the BIBB Error which includes the message number, object identifier and message "no such object."

Services include events such as requesting and receiving values for real and virtual points and alarms; sending and receiving instructions to set data values, enable and disable points; setting up and distributing schedules and trends; setting up and collecting trends; sending and receiving proprietary information.

4. Network – Transporting Request and Responses

For BACnet devices to communicate, they must be connected to each other. This can be over a network or phone line and there are many options directly available. In general, faster networks are more expensive both in cabling and the transmitter/receiver.

Networks
Ethernet 10 and 100 Mbps on twisted pair cable, coaxial cable, fiber optic cable
ARCNET 2.5 Mbps
MS/TP (master-slave/token-passing) < 1 Mbps, often 76.8 kbps, twisted pair wiring
LonTalk by Echelon
Phone line
PTP for phone lines or hardwired EIA-232 wire
Internet
BACnet/IP using Broadcast Management Device (BBMD) for internet communication

The use of the internet opens up many new possibilities and the BACnet protocol is being extended to make full use of the internet, including the transmission of web pages. A simple example of the use of the internet is in multi-tenant buildings. Tenants can be provided with access to their part of the DDC system without the building system having to be physically wired to any tenant computer. This avoids both physical cable coordination with tenants and network security issues. Within the building it enables staff to have a virtual thermostat – an image of a thermostat which shows the current temperature and set point, plus allowing them to raise or lower the set point within preset limits. Internet access can also allow tenants with space in several buildings to have a single staff person access the DDC system in all their spaces.

The internet raises additional concerns about network security. Typically, a networked DDC system will have security software which requires a password for access. Larger systems will have a hierarchy of passwords where the highest level gives access to everything and allows any changes to be made. Lower levels of access provide more limited access. The day-to-day operator may only have access to limited points and only be able to make simple adjustments including on/off control. Once access to the Internet is provided additional firewall security is needed.

A router can be used between any of the networks to pass on the BACnet packets in the changed transmission format. Thus, a large site could use Ethernet between buildings and to the operator workstation and MS/TP as the network (often called "field bus") around the building. In this situation each building would include an Ethernet to MS/TP router. Then, when an extension was added, the new vendor used ARCNET for their network requiring an Ethernet to ARCNET router as shown in *Figure 11-16*.

The use of a gateway is common for large equipment, including boilers and chillers. The chiller manufacturer may have its own proprietary control panel. A gateway can then transfer the necessary information from the chiller system into BACnet for transmission to BACnet devices. Typically, a limited menu of data is needed to go each way. From the operator to the chiller might be "turn on," "turn off," "set chilled water supply temperature to °F," "limit capacity to

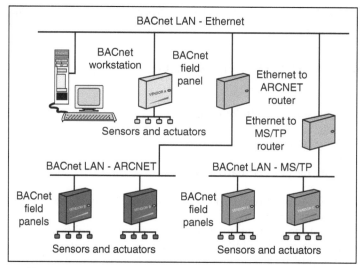

Figure 11-16 BACnet System with Various Network Protocols

x%," while the chiller might report "I'm on," "I'm off," "chilled water leaving temperature," "% load," "trouble alarm," "lockout." This enables the operator to control the chiller but it protects the chiller system from any software alteration or adjustment. In many situations this is an excellent equipment safety feature!

In the same way, a boiler manufacturer may have standardized on using an open standard protocol Modbus. This may well be ideal for the powerhouse with boilers, variable speed drives, and main electrical switchgear all communicating using Modubus. The remainder of the site may be using another protocol with a gateway between the two.

11.5 LonWorks

LonWorks® is a data communications technology based around the Neuron® chip. The Neuron chip is a very large scale-integrated (VLSI) chip. It includes the software to provide the communications protocol and point data. The communications protocol is called LonTalk. The Neuron chip provide includes the LonTalk protocol, the ability to communicate with input and output devices, and the ability to run user application programs. The basic LonWorks system building blocks are:

- LonTalk® communications protocol – what data and what format to pass information on the network.
- LonWorks® transceivers – the interface between the Neuron chip and the network cable.
- Development tools, including NodeBuilder® for developing the software in a Neuron chip, and LonBuilder® for developing custom devices and building a network.

These products are proprietary and are made by numerous manufacturers, and the technology has largely become an ANSI Standard. The name starts with Lon, meaning Local Operating Network, a network dealing with operations. This neatly differentiates its focus from LAN, Local Area Network, which focus on all types of communication in a building or organization. Lon technology is aimed at the efficient flow of control data.

Each node on the network contains three components: transceiver, Neuron chip, and I/O circuits. Each Neuron chip has a number of defined Standard Network Variable Types (SNVT), and each type contains a network variable (nv) value. This network variable may be an input, nvi, or output, nvo. For example, the SNVT_temp contains the nvo value temperature in degrees centigrade. The network variables for HVAC are mostly dealing with the HVAC issues, but a few are needed to establish network behavior. *Figure 11-18* shows the structure of one node with just two of many network variables: temperature input and heater output. A VAV node, for example, has eight mandatory and 24 optional variables.

Each nv has a rigorously defined format. Temperature, for example, is defined as being between -273.17 and $+327.66°C$ with a resolution of $0.01°C$. When a temperature is transmitted it is binary coded within this range, sent and decoded by another node. If one wants to display a temperature in $°F$, then it must be converted from $°C$ before displaying. This level of formality is chosen to minimize program size and message size. Thus, sending a temperature from one temperature output to a temperature input only requires sending the number representing the temperature.

A range of published standard Functional Profiles define all the Network Variables in the node. These profiles range from the simple temperature sensor to roof top unit controller in the HVAC field as well as lighting, smoke/fire detection, and specialized industrial profiles.

The standardization around the Neuron chip and setup equipment makes for relatively easy entry into the small scale DDC controls business. The NodeBuilder is used to program each node. Then, LonBuilder is used to model and configure the network. Including a Network Service Interface module allows an operator workstation to be included. For a small system it is relatively simple. However, larger DDC systems are more easily tested and configured with manufacturer specific development tools. Here the issue of interoperability becomes a challenge as a system installed by one supplier may have to be completely reprogrammed for another supplier to make additions or changes.

The development and control of the technology is largely controlled by LonMark International (LMI) and LonMark Americas (LMA), nonprofit organizations with world wide membership of interested parties who develop and maintain the interoperability rules and guidelines for LonWorks application. LMI operates a testing and certification program for devices. Details of tested devices, manufacturers and all about LonWorks are available at www.lonmark.org

Although BACnet and LonWorks are not compatible, the LonTalk network protocol may be used to transport BACnet packets of information between BACnet devices. This does not enable BACnet devices to communicate with a standard LonMark certified device. A gateway would be required to do the necessary translation between the dissimilar data protocols.

In the hypothetical example in *Figure 11-17*, there are two LonMark and two BACnet devices on the same LonTalk network. The two LonMark devices can

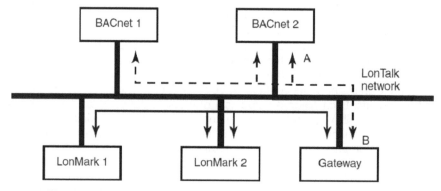

Figure 11-17   BACnet and LonTalk Controllers on a LonTalk Network

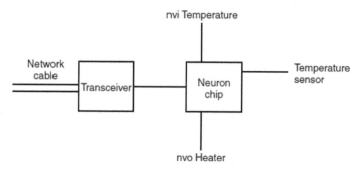

Figure 11-18   Outline of a Single LonWork Node

communicate between themselves. The two BACnet devices can communicate between themselves. For the LonMark 2 device to communicate with the BACnet 1 device a gateway is required. It receives the message from the LonMark 2 device and translates it before sending it to the BACnet 1 device. The network protocol used by the gateway is LonTalk in and LonTalk out. What the gateway does is translate the LonMark data format into BACnet data format.

The Next Step

The final chapter discusses the issues around the benefits and challenges of DDC systems. There are many advantages that are not available from any other type of system. These arise from the ability of a DDC system to store and manipulate data and to display it at any internet connection in the world. Deciding which of these benefits is to be included and ensuring that they work is is a challenge for the DDC system designer.

Bibliography

See the end of Chapter 12 for a list of DDC resources and bibliography.

Digital Controls Specification

Contents of Chapter 12

Study Objectives of Chapter 12

In the previous two chapters, we considered hardware, software, networks, and the issue of interoperability. In this final chapter on DDC, the issues of specification will be the focus. The specification is the document that defines for the contractor what is required. It is also the document that is used to check that the contractor has completed its contract and that performance is being met. To effectively achieve these objectives the specification must be clear and unambiguous.

To write a specification that is clear and unambiguous you need to know your objectives. Establishing the objectives is always dependent on what the client wants and is willing to pay for in both the initial construction and also in the ongoing operation and maintenance.

This chapter starts with discussion of what the client wants and goes on to discus some choices to be made in specifying a system. For specific, detailed, guidance on writing a DDC specification, including sample text, the ASHRAE GPC-13-2007 Guideline to Specifying DDC Systems is a very useful document. This guide provides advice and sample specification text based on BACnet but it can easily be modified to another protocol.

After studying this chapter, you should be able to:

Understand the benefits and challenges of DDC systems and how adding benefits may well add challenges.

Be aware of the issues of interoperability when choosing and bidding DDC systems.

Have an understanding of some of the possibilities of monitoring performance of DDC systems.

Be aware of the electrical issues that arise in DDC systems.

Know where to look on the internet for more information on DDC systems.

12.1 Benefits and Challenges of DDC

DDC systems can be extremely sophisticated and perform control logic that was quite impossible with any previous system. Unfortunately, many owners do not have operating and maintenance staff that are trained to understand and utilize the available potential of DDC. In addition, owner's staff often do not understand even the simplest ways of monitoring system energy performance. As a result, the anticipated control performance may not be met and the energy consumption may be much higher than expected or is necessary.

To deal as effectively as possible with this issue, a designer must start a project by finding out the owner's business motivation, and likely operation and maintenance abilities. The initial owner may be a developer who is going to sell the building and is not particularly concerned about performance or energy costs. At the other extreme are the owners who are building a performance and energy showpiece for their head office. In the first case, a simple system with little sophistication will likely suit the situation. In the second case it will be valuable to work up the project objectives with the owner, the architect, and electrical designer so that there is a common, agreed set of project objectives and strategies.

This cooperative approach will help use a systems approach to design rather than the (oversimplified) situation where the architect designs the building and then mechanical and electrical designers prepare their design. By systems design we mean considering the building as a whole system and designing the architecture, mechanical and electrical, to be efficient rather than considering each component separately. A mechanical example for a cooler climate is putting funds into high performance windows and omitting perimeter heating. This example does not produce a more complex system for operation and maintenance.

A combined mechanical and electrical example is using occupancy sensors to switch off lights and to provide a signal to the HVAC system to reduce outside air and allow temperatures change to the unoccupied limits. Although this approach has the potential to reduce lighting and HVAC energy consumption, it also raises the complexity of the HVAC DDC controls. Is the owner committed to providing the resources to run such a system effectively?

When first discussing the DDC system with an owner it is useful to have a clear understanding of the benefits and challenges of DDC. The following set of nine benefits is taken from the ASHRAE GPC-13-2007 Guideline to Specifying DDC Systems, with additional comments.

1. DDC systems can reduce energy costs by enabling mechanical systems to operate at peak efficiency. Equipment can be scheduled to run only when

required and therefore generate only the required capacity at any time. Additional savings are possible if the DDC system is used for more sophisticated purposes than timeclocks or conventional controls. If the DDC system simply duplicates the function of these devices, there may not be a significant reduction in energy consumption.

"Operate at peak efficiency" what does this mean? How will the client know what the initial efficiency is and what it is in a month, a year, a decade? Efficiency is generally defined as useful output divided by total inputs. In HVAC systems "useful output" is keeping a process or people in the required environmental conditions. Unfortunately, we cannot measure comfort and are limited to measuring cooling output, which is not an entirely satisfactory metric. We will consider the efficiency of a chilled water plant in a later section and how that may be monitored.

When designing a system think about "What has to be provided when?" In addition to time scheduling, there is the issue of "How much must be provided?" Simple examples include adjusting the outside air volume to match actual occupancy needs and adjusting duct static pressures down so as to provide only the required airflow.

Even scheduled running hours can often be significantly reduced if occupants can turn on the system for their area for a limited time (2 h, for example) when they are present earlier, or later, than the core operating hours.

Every watt of lighting in hot weather adds to the cooling load, so cooperating with the lighting designer to minimize lighting loads and lighting on time also reduce cooling loads. People are generally poor at turning off unnecessary lighting, so automating switch-off can produce lighting power savings of 50% or more.

2. DDC systems have extensive functionality that permit the technology to be used in diverse applications, such as commercial HVAC, surgical suites, and laboratory clean rooms. For example, they can be used to measure the amount of air delivered to each area, compare this to the ventilation needs of the building, and then vary the amount of outdoor air introduced to meet the ventilation requirements of ANSI/ASHRAE Standard 62-2007. This level of control is not practical with pneumatic or electronic controls.

   The level of additional functionality available through the use of DDC often requires a more carefully designed HVAC system and better trained staff to effectively operate it. Be very careful not to over design for the staffing capabilities, and to ensure that the control logic is clear to the operating staff.

   Note that there is considerable scope for automatic fault detection in DDC systems but it is not for free. Manufacturers and independent organizations are now offering fault detection programs and independent data analysis tools that will become more readily available and easier to implement in the future. Software can be provided on the system or the system can be interrogated over the internet and the analysis and reports generated at the office of the service provider.

3. A DDC system that controls HVAC systems in commercial, institutional, and multi-family residential buildings will provide tighter control over the building systems. This means that temperatures can be controlled

more accurately, and system abnormalities can be identified and corrected before they become serious (e.g., equipment failure or dealing with occupant complaints).

A DDC system cannot overcome system design problems. Under capacity, poor airflow in occupied zones, and an inability to remove moisture without overcooling are design flaws not control flaws. No additional DDC sophistication will resolve these issues. Be very careful to avoid blaming the DDC controls until it is clear that the system has the capacity to perform. This will avoid wasting considerable resources, making the DDC system seem unsatisfactory, and generally frustrating all parties.

In addition to providing tighter control, DDC can also be used to provide occupant control not practical with other systems. For example, providing occupant adjustable thermostats is a real challenge in many buildings as the occupants make extreme changes. With a DDC system, the thermostat limits can be individually set with a narrow band, say 73–78°F in summer. This prevents anyone turning the thermostat down to 65°F or even lower and having the cooling running flat out to produce a lower-than-necessary temperature and excessive energy consumption.

Systems with internet access can be set up so that occupants can adjust, within limits, the temperature in their space using their own PC and standard internet browser. This may be considered as a valuable feature to the owner of a high end office tower with demanding tenants, or as a complete waste of money by another owner.

4. In addition to commercial HVAC control, DDC can be used to control or monitor elevators and other building systems including fire alarm, security, and lighting. This enhances the ability of maintenance staff/ building operators to monitor these systems.

Note that the speed of response and circuit monitoring usually requires specialist DDC equipment to control non-HVAC systems. Thus, the controller running an HVAC plant may well be a different controller than the controller running the security system in that area. Both controllers may use the same data and communication protocols on the same network and appear as the same for the operator at their workstation.

More and more frequently the security, fire, and other systems provide inputs to the DDC system, even if they are not integral with the building HVAC system.

Several HVAC controls manufacturers are now providing lighting and access control. Note that open bidding becomes more difficult with increasing integration of disciplines, and requirements for all areas must be very clearly identified for getting bids. This integration is moving towards a Cybernetic Building System (CBS) which involves intelligent control, operation, and reporting of performance and problems. The capability exists with DDC systems to include self-commissioning. Examples include controllers which retune parameters as conditions change. A useful example is on a large air-handling unit, altering PID loop parameters when changing from controlling the heating coil in winter to cooling coil in summer.

5. DDC allows the user to perform intricate scheduling and collect alarms and trend data for troubleshooting problems. The alarm and trend

features permit the operator to learn the heartbeat of the system. Observing how a mechanical system performs under different load conditions allows the programming to be fine-tuned and permits the operator to anticipate problems. This allows the occupants' comfort concerns to be dealt with promptly before the problems become serious.

The ability to collect trend data is particularly valuable both in setting up the system to operate but also in ongoing checking on performance. For ongoing checking, some standard trends and reports should be established and included in the installation contract requirements.

6. Many DDC systems can provide programming and graphics that allow the system to serve as the building documentation for the operator. This is valuable since paper copies of the system documentation are often misplaced.

    The graphic display of the systems with current operating data is a major advantage of DDC. On larger systems it is very important that the hierarchy of displays and their format makes it easy for the staff to navigate quickly and reliably to what they want displayed.

7. A DDC system requires significantly less maintenance than pneumatic controls. Pneumatic controls need regular maintenance and must be periodically recalibrated. The use of DDC results in lower preventive maintenance costs due to calibration and also lower repair costs for replacement of pneumatic or electromechanical devices that degrade over time. All systems require some maintenance. Sensors such as relative humidity and pressure sensors require regular calibration regardless of what type of system they serve.

    The actuators and mechanical parts of the system, valves and dampers, will still need maintenance and will still fail. Automatic detection of failure is particularly valuable for these components.

8. DDC systems reduce labor costs through remote monitoring and troubleshooting. Paying a technician to drive to the building to deal with every problem can be minimized with DDC. In many cases, on-site operations can be eliminated or reduced to a single shift. Problems with the building are called out to a central monitoring service or to a technician with a pager, and often may be rectified from remote locations by modem.

    External monitoring and maintenance can be contracted to an external organization. The monitoring can be done from anywhere in the world where internet service is available, although the hands-on maintenance will need to be from a local provider.

9. DDC systems are programmable devices. As the needs of a building change, the system can be reprogrammed to meet the new requirements.

    This is both an advantage and, unfortunately, a disadvantage. Unless the operating staff really understand and agree with the control processes they can, and do, modify things. Often the result is very poor building performance.

    One of the advantages of having the system formally commissioned (having a separate commissioning agent test and verify the correct installation and performance of the system) is that there is a clear definition of how the system was operating when installed. This knowledge makes later checking on performance much easier as one has verified initial operation.

12.2 Design

Your specification will usually be written after the supporting system diagrams and points lists have been completed. Start the system design by dividing the project into its main systems, typically: air handler plus associated zone controls, boilers, and chilled water system. For each system work up the control sequence of operations for:

normal operations, including: unoccupied, startup for occupied on schedule, startup for occupied outside schedule, summer, winter, and load shedding.

failures, including: freeze protection, excess humidity, network communication lost, power failure, and restart.

special situations, including: fire/smoke control, fireman operation, and access triggered events.

data collection: specific trend logs, running hours, and energy usage.

Note that, in some situations, it is simpler and less confusing to complete the HVAC system controls first. Then consider the special situations separately and their overriding effect on the individual HVAC systems controls, as indicated in *Figure 12-1*. As an example, when a fire detector goes into alarm in System 1 what should happen in Systems 1, 2, and 3, the lighting controls (all means of escape routes lighting on), and access system?

In many buildings, there are several almost identical systems. In this case the sequence of operations for a base system can be produced with additions/deletions for each of the other similar systems. Be very clear about which variations apply to which systems to minimize confusion and mistakes.

Based on the sequence of operations, a hardware (analog input, AI, analog output AO, binary input BI, binary output BO) and software points list is produced. This list should identify whether the points are to be trended and whether the trend is to be based on a time interval (every 15 min), change of value (fan on to off), or amount of change in measured variable (±5% relative humidity). The points list is also a convenient place to identify which points are to be shown on the operator workstation graphics. Many of the software points known at this time will be alarms and these may be shown on the

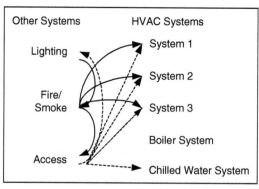

Figure 12-1   Superimposing Control Requirements from Other Systems

graphics as well as announced as alarms. For example, a high mixed air temperature alarm could require the mixed air temperature to flash on the screen. Some designers also choose to make up a numbered schedule of graphics for the operator workstation. The points list can then have the correct graphic identifiers against relevant points.

The points list may also be used to identify the sensors and actuators. If a numbered schedule of sensors and actuators is produced, then the number can be included as another column in the points list. Other columns on the points list may identify incoming and outgoing points for other systems such as lighting and access and if they are provided by other contractors such as electrical or plumbing.

At this stage the designer really knows the system and how it is expected to work. This is a good time to work up the particular requirements for the operator and maintenance workstations. Typically, there are four functions to be addressed:

- Data storage – where will long term trends be stored and backed up to secure storage media?
- Software program location – where is the system software for reloading and modifying the system?
- Operator workstation – where is the day-to-day monitoring of the system?
- Maintenance access – how will the maintenance person access the system?

Many systems have a single fixed operator workstation which deals with the first three items. This single workstation may be connected into a controller or into the main network. For maintenance, a portable device with an LCD or LED character display, or more commonly a laptop, can be plugged into any controller.

Depending on the system, the portable maintenance unit may be plugged into end devices such as thermostats. This is a very convenient feature as it allows the maintenance person to make changes while they are in the controlled space or close to the controlled device.

On larger systems, most often with an Ethernet network, the system may have several workstations. Not all of these are necessarily dedicated to, or part of, the purchased system. For example, a large campus of several buildings with on-site maintenance staff might have a fixed workstation in the boiler/ chiller plant for day-to-day monitoring. The controls shop could have the workstation which has data storage and contains all the software for reloading and system modification. The site energy manager may have a PC which is connected to the HVAC network, and is used obtain data from the system and monitor utility usage. This PC would typically not be part of the HVAC system; it would only have access.

Note that, if the system uses the site Ethernet network, any PC on the network can connect to the system. This open connectivity makes security particularly important. The HVAC system must have password security and passwords should be carefully controlled. The hierarchy of access – who can access what information and make what changes – must be carefully thought out. In general, it is wise to restrict the ability to make any program changes to well trained personnel as it is easy to make a minor change and completely mess up the program operation.

One way to minimize the temptation to make changes to the system is to provide clear, comprehensive, and informative graphics. Graphic packages from different manufacturers can vary in complexity, availability of 'canned' images, colors, speed, compatibility with common graphic generation tools such as Paint and Adobe, integration, colors, animation, and connectivity programs such as PCAnywhere, Laplink and VNC (trademarked products). Many packages include a library of component images which can be assembled to provide a cutaway view of the system, or part of the system, as shown in *Figures 12-2* and *12-3*. Most graphical software packages also allow for bitmap or JPEG formatted file-based photographs to be used. The photo image of the equipment or device is placed in the background of the graphic screen, and then the real-time data are shown in the foreground.

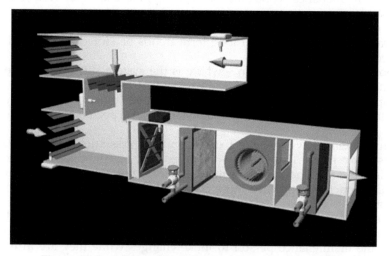

Figure 12-2   3-Dimensional Cutaway Image of an Air Handler

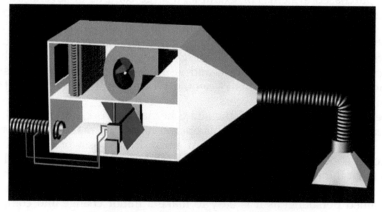

Figure 12-3   Graphic Image of a VAV Box

Note that if the system has high speed internet access then some, or all, operator workstation functions with graphics can be anywhere in the world that there is high speed internet access. This provides the possibility of having all the monitoring done at a remote location, with local maintenance staff being dispatched to deal with problems and routine maintenance. However, if dial up or low speed internet access is used, at either the system or remote location, the graphics will slow the system response so as to be unacceptable.

Typically, the beginning of the graphic displays starts with an Introduction or Table of Contents menu or Map of the area. Then it can show a "master" listing of mechanical-electrical systems, and, sometimes, simplified floor plans of the building. For example, this might include each chilled water system, hot water system, chiller, boiler, pump, tower, valve, air handler, exhaust fan, lighting system, and all terminal equipment. This "master" page can have icons that will automatically transfer to that graphic for the equipment being monitored. Point information on the graphic displays should dynamically update. The point information should show on each graphic all input and output points for the system, and also show relevant calculated points such as set points, effective set points after resets, occupied status, applied signals such as voltage or milliamps, positions, ambient conditions, alarms, warnings, etc.

The time for graphics to show and include current data for the data points should be limited to 10 s. This may limit the number of data points to, say, 20, which is quite adequate for most graphics. Note that "current data" can be interpreted in three ways:

1. Data already in the operator terminal. The age will depend on how frequently the operator workstation polls for data.
2. Data collected from the relevant master controller. If this master controller collects data by sequentially polling slave controllers, the data may be as much as a minute old.
3. Data collected from the actual data points.

Once the graphic is up on screen, the data should be regularly refreshed every 8–10 s. The updated data should be current values collected from the data points within the last 6 s – not the most recent of a polled data set in a master controller. The requirements for operator terminal speed must be clearly identified in the specification, as speeding up the system after completion may be impossible without effectively rebuilding it.

It is also helpful to have maintenance information that can be fed from the graphics page for each control system. For instance, an icon at the valve, when pressed, would open up the technical data sheet and 'Maintenance & Operation' page for that valve.

Global points, such as outside air temperature and humidity, time, date, alarms, and occupied status, should be displayed on each appropriate graphic.

An important function of the control system and associated controllers and workstations is that of notifying operating personnel of alarms, warnings, advisories, safety faults, or abnormal operating conditions [11]. Notification methods include visual and audible alarms, printout of alarm conditions with date and time indicated, and automatic telephone calls, via modem or the internet, to off-duty staff or outside service companies. The DDC system can be programmed to respond in specific ways to specific alarm conditions, including shutting down equipment, starting back-up equipment, adjusting set points, initiating smoke control modes, and other appropriate measures determined by the control designer. Modern performance evaluation indices can be used to compare the performance of similar controllers and identify faulty controllers more effectively. Custom alarm messages can be generated for different types of alarms, telling the user what to do in the case of that particular alarm. For example, if the high-pressure static pressure alarm goes off in the AHU, an alarm message can come up on the screen and tell the user what to do and how to fix it, in addition to activating a graphic that will show them where it is and how to find it.

Alarms come up on the screen of the operating system that let the operator know of the alarm condition. Note that alarms may not come up quickly; Guideline 13 suggests a time limit of 45 s. This is adequate for many HVAC situations but life safety and process systems often require immediate, or almost immediate, notification. If the operator password level is low (restricted access to view and alter the system), the alarm message and its instructions can be viewed but cannot be "acknowledged" it so that it goes away on the screen. Higher-level password operators can come in later and review the system actions and alarms, so that a designated responsible person can be made aware of them (*Table 12-1*). This feature is important, in order that alarms are dealt with and not ignored.

The control system software and front-end computer can also provide maintenance scheduling. It can be manufacturer provided or third party. Programs can be provided that will print maintenance work orders and follow-up reports on a regular basis, or, after a certain number of equipment run-hours, an alarm triggers or there are timed routines, thereby allowing improved maintenance and performance of the equipment. ASHRAE Standard 180 *Standard Practice for Inspection and Maintenance of HVAC Systems* will, when published, make several recommendations on using DDC systems for maintenance and operation of a typical building. Typical examples include filter changing, coil cleaning, and routine lubrication of motors at run time or time intervals. These are just a few examples of how proper care and maintenance of HVAC&R systems can improve energy efficiency and functionality.

The users and programmers of the control system use system diagnostics and testing routines extensively. Each manufacturer's setup is unique, but its intent is to foresee hardware and software problems before or as they occur Seem *et al.* (1999). The diagnostics software creates alarm messages and warns the user of the problem it has detected. The testing program allows for the manual testing of all inputs, outputs, and loops in the DDC control system. It is beneficial for initial testing and troubleshooting, but, as the life of the system goes on, it serves a valuable and useful tool for seeing the system respond to artificially created variables and situations. Some DDC systems also have a "live demo" testing software available, so that most of the DDC system can be seen operating by a test database of values operating in simulation of real-time.

Help screens and training venues are available from most DDC systems. The software documentation should be in hard copy as well as in software so that the operator can access it regularly from the PC. These help and training venues should also be updated periodically by accessing to the manufacturers

Table 12-1   Alarm Printout

Alarm	Status	Time	Event Class	Message
ESB AH3 SF STS	Normal		Notification	
Transition	Normal	13:17:30 19 November 2007		AH3 SUPPLY FAN NORMAL
Transition	Alarm	13:14:36 19 November 2007		AH3 SUPPLY FAN FAILED TO START
CTC AH3 SF AL	Alarm		Critical	
Transition	Alarm	04:50:34 20 November 2007		CTC AH3 SF STS (off) does not match CTC AH3 SF (start) AH3 fan failed
UCT RHT P ALM	Normal		Critical	
Transition	Off-alarm	09:08:27 19 November 2007		CHECK UCT REHEAT PUMP; REPORT PROBLEMS
POW FREON ALM	Normal		Critical	
Transition	On-alarm	11:42:17 19 November 2007		Freon levels in chilled water plant too high
DUF RAD RHT ALM	Low-limit			
Transition	Low-limit	11:57:52 20 November 2007	Maintenance	Please check Rad Rht pressure: report problem

web page or as automatic downloads. Care should be taken not to download manuals for software versions later than the one running the system.

Keeping backups of the programs and data is very important. The programs are the software used to run the system, make changes, produce graphics, and other add-ons such as maintenance scheduling. Copies of these programs should be stored carefully, not just "on-the-shelf" with other manuals. Remember that staff turnover occurs – not always under amicable circumstances. Keeping a current copy of the controller programs and historical data may prove challenging as it is constantly changing. One method is to have the controller programs and historical data automatically backup to a separate hard drive. Then, on, say, a weekly basis the backup can be copied to a CD/DVD and stored elsewhere.

The two questions relating to backups that need to be answered are: if everything is on one operator workstation and the hard drive fails how will the system be restored? If there is a fire in the room with the operator workstation and backup hard drive, how will the system be restored? To ensure that these two questions can be answered satisfactorily you could specify that the process for both situations be taught to the operating staff so that they can demonstrate the recovery process by doing it themselves. One client defined their requirement as follows:

Software, backup facilities, and permanent instructions shall be provided so that on total loss of the operator workstation an equivalent new PC with Windows XP can be reloaded with all operating and design software, the previous day's database, and all archived history. The newly loaded PC is to operate the same as the lost machine.

Now, depending on whether it is a new system in a separate building or an addition to an existing DDC system, the specification can be written. Many designers already have a specification and it is a matter of modifying it for the specific project. If, however, the specification is to be written from scratch, then assistance should be sought. An obvious, and often used, source is a controls vendor. Be aware that the sample will be written to suit the vendor's system. This is not a criticism but it is a warning. Read it carefully and decide whether the degree of specification suits what you require in this particular project. It will likely be based on a particular control protocol. Is it the one you require or would others be acceptable, or even preferable? Is the wiring specification overly restrictive, inadequate, or unsuitable, for other vendors? Does it suitably specify a point naming convention, spare point capacities, operator terminal features?

Two other sources are recommended. The first, already mentioned, is ASHRAE Guideline 13-2007 Guideline to Specifying DDC Systems which includes sample specification language for much of the specification. It is a great help but again read it carefully and decide what you want for your system. The Standard uses BACnet as its basis for control protocol. Is this what your project needs? It is somewhat brief in content on commissioning, warranty, and backup requirements so again, what do you want for your project?

The last suggested source is based on Guideline 13 and is an online site which produces controls specifications for specific systems, https://www.ctrlspecbuilder.com/. One chooses the system type from the given choices and then makes a series of choices about the system from a preset menu. The site is easy to use and produces schematics, in VISIO (and can be imported into AutoCAD), as well as specifications. This site is supported by a controls vendor but the controls specifications it produces are based on ASHRAE Guideline 13 and do not contain any proprietary product specifications. Go and experiment with it. To repeat the warning: this is another good source but do read through the specification it produces carefully, and make sure the specification works for your project, and modify it as needed.

If your DDC project must now, or in the future, operate with another system, the issue of interoperability must be addressed. Including text such as "must be BACnet compatible" or "must use the same operator terminal as the existing system" does not help the bidders decide what to provide. Neither does it help you decide whether what is provided does what you require.

If connection to an existing system is required, discussion with potential bidders can be extremely valuable for specifying a suitable solution. Which brings us to the next section on bidding and interoperability.

12.3 Bidding and Interoperability

Bidding processes are often constrained by the client and may also be constrained by the funding organization. Find out these constraints before writing your specification. Note that they may be formal, such as a state requirement that open bidding is required for all state buildings. Requirements may also be unwritten and due to prior experience with a product or supplier. Ideally, the list of acceptable bidders for DDC should be decided with the client and included in the specification to avoid having to assess company capabilities and compatibilities under the time demands of the bidding process.

Guideline 13 includes a list of issues that can be used to pre-screen vendors. Prescreening should be done with the capabilities of the client's operating and maintenance staff in mind. With very competent staff resources, local support for the system may not be critical. On the other hand, if local support from the vendor is critical, bids for installation may also include maintenance contract options.

On larger projects it may well be advisable to include one, or more, members of the client's staff in this decision process. A staff person from the maintenance department can do a lot to increase the department's commitment to making the project a success. Having a staff person from the purchasing department involved during the selection process can provide valuable support during the bidding process, as he or she will have some understanding of the choices around the DDC system being purchased. He or she may also understand that lowest-first cost from the controls vendor is not always best long term value.

If the DDC system is to use the business LAN, having a representative from the IT department involved can also make for a smoother integration process. Start talking to the IT department really early on in the project. Do note that device names and terminology are not consistent between the IT world and HVAC world! Having the parties describe what a device does can often lower the tensions, as terminology gets sorted out. Agreement is needed on who will be responsible for network issues on the shared network. If there is a corporate overhead cost distribution for the network, this needs to be within the building prime cost and operation budgets. Remember the network must be up and running for the DDC system to work. If the system is down for an hour each week for IT backup reasons, this may be the reason for having a separate network for at least priority parts of the HVAC system. As the quality of service matters much more in some places than others, these out-of-hours periods of downtime may not be significant or they may be critical. If possible, have these issues settled, or at least clearly identified, before facing vendors.

The question of integration brings us back to the general issue of interoperability, the ability of parts of the DDC system to work together. If the project is stand-alone HVAC and not expected to be changed or expanded then a single vendor system has only to work with major equipment such as chillers and boilers. If the system is to be connected to an existing system or is expected to be enlarged over time, interoperability merits careful consideration. Remember the basic issues around interoperability and the use of open protocols. The following are the positive ones:

- ability to use different vendors for different parts of the HVAC system plus additions, and modifications
- integration with other applications such as access, elevators, lighting, and life safety having a single operator interface for the entire system
- ability to share the organization business network cabling
- investment protection through a longer system life, as parts can be upgraded and replaced.

There are also some challenging issues to deal with even when an open protocol is being used. Initial integration of the different systems is not plug-and-play, and specifications must be clear about which vendor is taking responsibility for initial overall operation. If the specification, as is typical, requires free software updates during the warranty period, responsibility for getting these installed and fully working should also be specifically specified for one vendor.

Once installed, maintenance of each vendor's system is required. This may be in-house, which will require additional staff training time, and additional maintenance tools and spare parts, or additional outside maintenance contracts. Note that the use of an open protocol does not mean that the same maintenance tools will work for parts installed by different vendors. Many vendors have designed their own ways of binding (interconnecting) LonWorks devices on a network. As an example, Vendor A's system tools may not be able to add a LonWorks device to the Vendor B network. The device is not the problem, it is how it is programmed to work on the network. Unfortunately, the full cost and time implications of maintaining multiple vendor systems is often seriously underestimated or ignored during the initial purchase decision period.

When designing a new system, if it is to work with an existing system with a single operator workstation, the first issue to be resolved is what gateways the existing system manufacturer has to other HVAC protocols. Since considerable data are likely to be passing at times, a high speed network protocol is also desirable. If none exists it is possible to get one developed for a large project but this route is prone to financial and time challenges. It may be better to accept the two operator workstations until the existing system can be replaced.

In the more usual situation, where at least one suitable gateway is manufactured, the next issue is which of the available protocols will be most suitable for the new system. If the system is to include connection to access, lighting, and fire/safety systems, this may be a challenge. Unfortunately, each industry is in the same situation as the HVAC industry with many existing proprietary protocols and open protocols gradually coming into general use and availability. To overcome this challenge, many of the larger HVAC controls manufacturers are forming alliances with access, lighting, and life safety system manufacturers to provide an integrated package.

Having resolved the issue of protocols at the system level, there is still the issue of protocols in the HVAC sub-networks. If the installed system is likely to be relatively static, proprietary sub-networks can be used. However, if significant changes or additions are likely it can be beneficial to require the system protocol to be used for sub-networks. The following is a simple example. Due to budget restrictions, several variable speed drives have been cut from the project. The client has asked for the controls point capacity and for the cabling to be arranged so that they could be installed at a later date. If a proprietary sub-network is installed the choice of variable speed drive may be limited when the funds are available to install them. If, instead, the system protocol had been required, this limitation would have been largely avoided.

12.4 Monitoring

One of the most useful features of DDC systems is the ability to monitor what is happening in the system. Seeing the system working and checking on performance is greatly assisted by recording, or trending, values of points at set intervals. For checking the performance of a control loop a short time interval, say 1 s, may be used. The minimum time interval and number of records will typically be determined by the controller and its memory capacity. For ongoing monitoring of system performance, longer time intervals, say 15 min, may be specified with the records being regularly uploaded and stored in the operator workstation. The points that are to be continuously trended to provide history should be identified in the points list.

Trended data can also be used to asses the comparative performance of the system over time and we will consider two examples: one for heating and one for cooling. Both of these examples involve relatively simple recording, data manipulation, and display. However, unless they are set up as part of the initial contract they may well be beyond the desires and capabilities of the accounts and operating staff.

Many building owners do not posses the staff competencies and incentives to monitor utility use and costs. Typically, the utility bills are paid from a utilities account by accounts payable staff that have no idea how the utilities were used. The maintenance and operations department are expected to maintain and operate the plant to everyone's satisfaction, with no downtime and an ever decreasing budget. With accounts payable lacking the ability, and the maintenance and operations lacking the incentive, it is no wonder that many organizations use, and thus pay for, more utilities than necessary.

Our heating example answers the question: "Taking into account the effects of weather, is this facility using more or less heating now than in the past?" Heating loads for the building fabric and air are proportional to the difference in temperature between inside and outside. If the temperature difference between inside and outside doubles, the conduction heat loss through walls, windows, and roots will also double. Similarly, the heat needed to raise outside air to indoor conditions will double.

The comparative effect of outside temperature changes can be estimated reasonably accurately by a simple process, often called the "degree day" method. If the outside temperature is recorded every hour for the (24 h) day, and the highest and lowest records averaged, one has a reasonable estimate of the average outside temperature for the day. The difference between this temperature and, say, 65°F gives a measure of the temperature-related heating load over the day. The temperature of 65°F is an arbitrary temperature base used by most environmental recording organizations in North America. Other temperatures are used in other parts of the world, 60°F, for example, in the UK. The temperature is chosen as the temperature at which buildings no longer require heating.

If the average were 42°F one day, the difference is $65 - 42 = 23°F$. This is called a 23-degree day, meaning that the heating load that day is the load that would occur if the temperature was 23 degrees below 65°F for the full 24 h. By summing the degree days for every day of the month, we get an estimate of how cold the month was relative to 65°F.

Occupancy will also affect the heating load. Day versus night and weekday versus weekend will have different heating requirements. However, over a whole month, these hour-by-hour variations usually average out so that we can compare one month with another month.

Figure 12-4 shows an office building plot of gas consumption against degree days for two years. The points lie closely clustered around the "best fit" line which is the baseline consumption for this building. If, in future months, points are consistently above the line, the building consumption is greater than baseline. In this case the points all lie very obviously on a straight line, but this is not always the case. A school closed over the winter break will show, or should show, a lower consumption in that month.

For a new building it will take months to build up the baseline, but over time the heating consumption can easily be monitored. If the utilities are not metered the quantities (not costs) from monthly utility bills can be manually extracted and used.

This example has dealt with answering the issue of monitoring heating energy versus heating load. Changes up, or down, can be due to system changes and external factors such as change in use of the building and additional energy conservation measures. In other situations, particularly where several pieces of plant interact, we may wish to monitor plant efficiency.

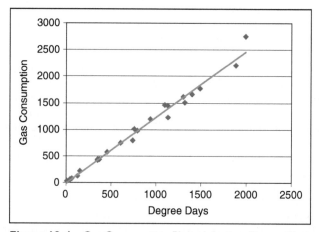

Figure 12-4   Gas Consumption Plotted Against Degree Days

As an example let us consider monitoring the performance of a chilled water system.

Many larger sites have central chilled water systems. Research has found that systems often run at well below anticipated performance and that the poor performance is often not known by the operating staff. To improve this situation it would be valuable to have the system report on chilled water plant performance. A typical electrically driven chiller system consists of four main components as shown in *Figure 12-5*.

- chilled water pumps to distribute the chilled water
- chillers to cool the chilled water and reject heat to the condenser water
- condenser water pumps to circulate the condenser water through the chiller and cooling towers
- cooling tower fans

The coefficient of performance of the plant is

$$\frac{\text{useful cooling produced}}{\text{chilled water plant power consumption}}$$

The coefficient of performance, COP, is dimensionless with the same unit on the top and bottom of the equation, Btus or kW. Alternatively the performance is often expressed in terms of kW input/ton of cooling output, kW/ton. A ton of cooling is 12,000 Btu/hr and a kWh is 3,412 Btu/hr. A good plant at peak efficiency may operate at 0.5 kW/ton.

$$\text{COP} = \frac{12,000}{0.5 \times 3413} = 7$$

Note that both ways of reporting are considering energy use and not energy cost. On plants which have storage, usually chilled water or ice storage,

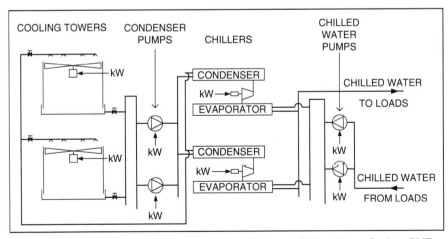

Figure 12–5   Chilled Water Plant, Power IN (kW) and Chiller Water Cooling OUT

minimizing cost, not energy consumption, is an objective. The concept is that it can be cost effective to use more low cost energy at night to produce and store chilling for the following day than produce the chilling at time of need. Because the performance of the plant is more complex to record and to present is not covered here.

In designing a monitoring system, the first, and most important, issue to decide upon is "what is the objective?" If the objective is to monitor the COP of the chilled water plant what information must be recorded (referring to *Figure 12-5*)? What has to be measured?

The obvious answer is all the kW inputs and the chilled water cooling – 8kW meters and a Btu meter for the chilled water. Obviously, but not necessarily. If the distribution panel can be laid out to supply the chilled water plant through a single metered cable or bus, as shown in *Figure 12-6* all power can be metered with a single meter. Perhaps the chillers are supplied at a higher voltage than the pumps then the question is can the chillers be on a single supply and the pumps on a single supply, using two meters instead of eight?

You may be wondering why one would consider meter consolidation when having the 8 meters provides information on how each component is working, not just the whole plant. The issue is what your objective is and what is the cost of meeting that objective. Predictably, the more accurate the meter, the higher the cost. If plant efficiency is the required information, then using just one input meter and one output meter only requires two accurate meters. It also has the advantage of reliability, as failure of either meter produces an obviously incorrect performance value.

Note that making choices about the metering will involve the electrical designer. One needs to think of the system as more than just the HVAC components.

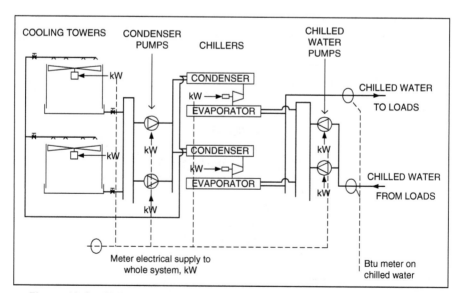

Figure 12-6 Chilled Water System with One Input Meter and One Output Meter

What accuracy do we need? ASHRAE Guideline 22-2008, suggests aiming at 5% overall accuracy. To achieve this we need to meter the electrical power and chilled water Btu. An electric meter with accuracy of $\pm 1\%$ is readily available, and $\pm 0.1\%$ is quite practical. Btu measurement is more of a challenge. The Btus are calculated from chilled water flow rate and temperature difference. The flow rate measurement is going to be dependent on both the quality of the meter but also the installed situation. A length of straight pipe before and after the meter is needed to achieve even flow, and this is often difficult to achieve. Note that it does not matter whether the flow meter is in the supply or return pipe or whether it is in the confines of the plant space or in a distribution main, as long as it is measuring the full chilled water flow. Be aware that factory calibration accuracy is rarely achived in the field, even with careful field calibration. Typically, $\pm 2\%$ accuracy can be achieved but $\pm 1\%$ is difficult.

Temperature measurement with a thermistor, calibrated over the narrow range of chilled water temperature (38–60°F), can provide an end-to-end (thermistor, transducer, controller) accuracy of 0.1°F. The differential temperature is the difference between two temperatures so on a system with a 10°F differential it is better than $\pm 2\%$, and close to $\pm 1\%$ with a differential of 15°F.

Combining accuracies to obtain an estimate of overall accuracy is beyond this text, but with better than 1% on electricity, and better than 2% on flow and tempertaure diufference, the target of $\pm 5\%$ can be met. Note that it requires considerable diligence and effort to achieve these accuracies with standard DDC control system components and the accuracy may drop seriously at lower loads. It can be beneficial to specify an energy management package from a specialist supplier with the outputs being connected to the DDC system.

On a system where we are measuring COP, how much energy is rejected from the system by the cooling towers and how would you measure it? Putting a Btu meter on the condenser water circuit is one way. However, flow meters on condensing water flows are often inaccurate due to the poor water quality. What heat is being rejected? The heat brought in by the chilled water and the heat put into the plant in the form of electrical energy, *Figure 12-7*, is the heat being rejected. There is no need to meter the heat rejection unless one is using it as a check on the COP input measurements.

For detailed information on assessing the efficiency of chilled water plants, refer to the proposed ASHRAE Guideline 22, Instrumentation for Monitoring Central Chilled Water Plant Efficiency. This standard dicusses the desing of chilled water systems and their metering, as well as providing example specification language.

Two useful resources for comparing building heating and cooling energy usage are ASHRAE Standard 105-2007 Standard Methods of Measuring, Expressing and Comparing Building Energy Performance and the US EPA Portfolio Manger. The standard provides standardized forms for data collection and comparison along with helpful conversions and energy values for fuels. The EPA Portfolio Manager provides US organizations with an easy-to-use online software that can provide comparative performance assessment and data for many of the common building types.

The systems performance and the success of these two metering examples depend on measurement accuracy and reliable recording over time.

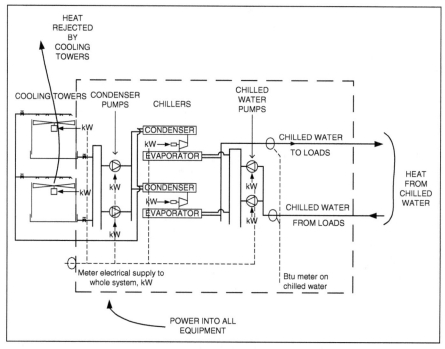

Figure 12-7   Energy Balance in Chilled Water Plant

The accuracy of metering depends on a chain of factors: sensor location, sensor accuracy, transmitter, cabling, and A/D converter. Typically, one specifies an end-to-end accuracy, as in *Table 12-1*, taken from ASHRAE Guideline 13 (Table 1).

Table 2 in the Guideline shows the requirements for control performance. It is a requirement that the control system maintain control of a variable within a specific range of values. Note that the two tables are independent. For example, Air pressure (ducts) ± 0.1 in. w.g. in *Table 12-1* is just adequately accurate to maintain the Table 2 requirement of ±0.2 in. w.g. However, the Air pressure (space) controller must be much more accurate than the reported value of ± 0.01 in. w.g. in *Table 12-2* to maintain the *Table 12-3* control requirement of ±0.01 in. w.g.

12.5 Wiring

Network Wiring

The performance of network wiring depends on the network layout, and the skills and diligence of the installers. The controls contractor must be made responsible for the complete network design and installation. The network design must conform to the manufacturers' wiring requirements, including the addition of controllers or other hardware. Design and specification wording do not produce a quality installation without competent installers.

Table 12-2 Reporting Accuracy

Measured Variable	Reported Accuracy
Space temperature	$\pm 1°F$
Ducted air	$\pm 1°F$
Outside air	$\pm 2°F$
Dew point	$\pm 3°F$
Water temperature	$\pm 1°F$
Delta-T	$\pm 0.25°F$
Relative humidity	$\pm 5\%$ RH
Water flow	$\pm 2\%$ of full scale
Airflow (terminal)	$\pm 10\%$ of full scale *(Note 1)*
Airflow (measuring stations)	$\pm 5\%$ of full scale
Airflow (pressurized spaces)	$\pm 3\%$ of full scale
Air pressure (ducts)	± 0.1 in. w.g.
Air pressure (space)	± 0.01 in. w.g
Water pressure	$\pm 2\%$ of full scale *(Note 2)*
Electrical (A, V, W, Power factor)	$\pm 1\%$ of reading *(Note 3)*
Carbon Monoxide (CO)	$\pm 5\%$ of reading
Carbon Dioxide (CO_2)	± 50 ppm

Note 1: Accuracy applies to 10%–100% of scale
Note 2: For both absolute and differential pressure
Note 3: Not including utility-supplied meters

Table 12-3 Control Stability and Accuracy

Controlled Variable	Control Accuracy	Range of Medium
Air pressure	± 0.2 in. w.g.	0–6 in. w.g.
	± 0.01 in. w.g.	−0.1 0.1 in. w.g.
Airflow	$\pm 10\%$ of full scale	
Space temperature	$\pm 2.0°F$	
Duct temperature	$\pm 3°F$	

This being so, specify that installers who are experienced in this type of installation are required to carry out the installation to the manufacturers' requirements. To avoid any "we didn't know" arguments, ensure that the controls contractor provides a copy of the relevant manufacturer's wiring requirements before any wiring is started.

This performance specification approach is far more effective than the designer attempting to specify the network and installation in detail. However, the designer may wish to include some specific requirements which may include:

- The network wiring be in a dedicated raceway, although this adds substantial cost.
- Spare cable requirements for future expansion needs.
- Labeling.

Effective labeling makes checking the initial installation and maintenance much easier and reduces mistakes. DDC sensors often terminate with two wires in an electrical box. When two sensors are side-by-side on an air-handling unit they both look the same. The boxes and wiring should be labeled. Although self-adhesive labels look more professional, the point names in permanent marker will still be there when the adhesive has failed.

Many control systems will have more than one network type. Typically, a high speed Ethernet with slower sub-networks. Specifications for both network types are required, and more experienced installers are needed for the high speed network. Be very clear about responsibilities, where the controls are to use the site LAN, and what testing and verification may have to be done to satisfy the in-house IT department.

Be aware of distance restrictions since, depending on the wiring configurations, the use of a signal booster repeater, may be necessary to maintain reliable signal strength.

Shielded wire may be used to limit the induction of noise onto the system's conductors. Manufacturers of control systems may or may not recommend its use. Some systems only require the shielding for communication wiring. When connecting the shield to ground it is generally desirable to only connect one end of the shield to its grounding source, and leave the other end not connected. It also should be noted that shielding can affect allowable wiring distances and installed cost.

Underground cabling can offer many other problems. First, the pulling of the cables into the underground conduit can be dangerous to the integrity of the network. The tension created by the installers can stretch the wire and cause the #18 AWG wire to become anywhere between a #18 AWG to a #24 AWG wire in the middle or at the ends, not to mention the obvious possibility of breakage of the wire totally. For this situation, using an "assisting pull wire" such as a #12 AWG for long pulls and wrap the network wiring around the assisting wire can relieve most of the tension.

Long underground pulls can create nicks or tears in the insulation, causing shorts and intermittent communication problems. Also, underground conduit will eventually get water in it. In any case, wet wire can mean a nightmare to the control system operator. All of a sudden the "no communication" light comes on and does not go off. Wire insulations can also be important; make sure the wire you are using is recommended for use in damp or wet locations. In general, the wire you would use in the ceilings, such as plenum or plastic jacketed, will not be appropriate for use underground.

When installing thermostats and wall mounted devices, it is recommended that an appropriate rough-in plate be used that will rigidly connect the plate to the structure and not allow movement. Furthermore, all surfaces should be kept level, and as close to fitting tolerances as possible. Cutting and patching should be kept to a minimum. The use of Molly bolts or drywall anchors for attaching wall devices is not acceptable for long-term reliability.

Air migrating from one part of the building to another may flow into the back of a thermostat, causing an incorrect temperature to be sensed. Depending on the height of the building, construction, and location, it may be wise to specify that the wall hole behind all thermostats be sealed against air leakage.

Fiber Optic Cable

Fiber optics is an alternative to a copper, wire-based network cable. A fiber optic cable consists of numerous glass fibers in a sheath. A transceiver, modem, at each end converts the incoming network signal to pulsed light signals using light emitting diodes, LEDs, which pass through the glass fibers to be read by the other transceiver and converted back to the network signal for onward transmission in a cooper wire (*Figure 12-8*). Fiber optic cables can communicate farther and faster than copper. The light signal is immune to electrical noise, ground potential differences, and lightning strikes, and is a good choice for use outdoors.

The drawback of fiber optics is that it has a significantly higher installed cost. A fiber optic installation requires more expensive transceivers, media, and connections. It should only be specified where the benefits justify the cost. For most DDC applications, the higher speed of fiber optics is not realized over that of the less expensive copper media. Fiber optic cable is available for lying in a 0.5 inch wide slot 1.5–2.5 inches deep cut in concrete or asphalt. The slot is cut with a standard diamond wheel, the cable inserted, and the slot sealed with hot asphalt or caulking in concrete. For building-to-building connection, where a copper wire cable route is not readily available, this can be an effective solution.

Fiber optic cable laying and connection must be carried out by an experienced specialist. Pulling tension, residual tension, and bending radii are very important, and the contractor must adhere to the manufacturer's requirements. Where a fiber optic cable goes directly to a workstation, it is advisable to terminate the fiber optic cable in a permanent location where it will not need to be moved during renovation or remodeling. The run from fiber optic termination to the workstation is by an Ethernet connection. This avoids any problems if the workstation has to be moved, as the fiber optic cable is fragile and easily broken.

Power Wiring

NFPA 70, also known in the USA as National Electrical Code (NEC), Article 770 should be followed. All line voltage power supply wiring, NEC Class1, will be covered by the main electrical specification for the project, which will be subject to local and national codes. Maximum allowable voltage for control

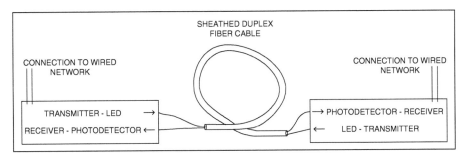

Figure 12-8   Fiber Optic Cable

wiring shall be the local user socket outlet voltage, 120 V in North America, and 110 up to 240 V in other parts of the world.

In general, power wiring should not be installed in the same relative area or raceway as the network wiring. A few guidelines below cover recommended installation practices:

- All low-voltage wiring shall meet NEC Class 2 requirements. Low-voltage power circuits shall be sub-fused when required to meet Class 2 current limit. (NEC Class 2 wiring is the circuit to a final device).
- Where NEC Class 2, current-limited, wires are in concealed and accessible locations, including ceiling return air plenums, approved cables not in raceway may be used provided that cables are UL/ CSA/ ETL or other code listed for the intended application. For example, cables used in ceiling plenums shall be listed specifically for that purpose.
- Do not install low voltage Class 2 wiring in raceway containing power Class 1 wiring. Boxes and panels containing high voltage wiring (90 V and above) and equipment may not be used for low voltage wiring (89 V and below), except for the purpose of interfacing the two such as for relays and transformers.
- Where plenum cables are used without raceway, they shall be supported from or anchored to structural members. Cables shall not be supported by or anchored to ductwork, electrical raceways, piping, or ceiling suspension systems.
- All wire-to-device connections shall be made at a screw-type terminal block or terminal strip. All wire-to-wire connections shall be at a terminal block. Note that wire nuts, soldered, and crimped connections should be avoided as they system testing and maintenance difficult.
- All wiring shall be installed as continuous lengths, with no splices permitted between termination points.
- Install plenum wiring in sleeves where it passes through walls and floors. Maintain fire rating at all penetrations using UL rated products.
- Use coded conductors throughout with conductors of different colors.
- Control and status relays are to be located in designated enclosures only. These enclosures include packaged equipment control panel enclosures unless they also contain Class 1 starters.
- The contractor shall terminate all control and/or interlock wiring and shall maintain updated, as-built, wiring diagrams with terminations identified at the job site.
- Post-installation residual cable tension shall be within cable manufacturer's specifications.

12.6 Commissioning and Warranty

Commissioning Process: a quality focused process for enhancing the delivery of a project. The process focuses upon verifying and documenting that the facility and all of its systems and assemblies are planned, designed, installed, tested, operated, and maintained to meet the Owner's Project Requirements (ASHRAE Guideline 0 The Commissioning Process).

Other definitions specifically require the provision of the system documentation and training being provided to the facility staff. Good commissioning goes much further than the HVAC systems contractor making the system work. Commissioning ideally starts with the commissioning agent making sure it understands the intent of the controls and their logic. During construction and on completion, random checks are carried out to detect any systematic errors and specific checks on control logic are also carried out. Finally, the commissioning agent ensures that the staff training is properly completed and documentation provided.

With the increasing complexity of modern buildings, and particularly the integration of lighting and other controls, commissioning is becoming even more important and valuable. Commissioning is an added cost to the initial project. Commissioning reduces the initial occupancy costs and disruptions as well as reducing ongoing utility costs. For some clients, minimizing occupant dissatisfaction can be more valuable than accountable financial savings, so the cost is secondary. Unfortunately, the initial cost and ongoing costs are separately accounted for in many organizations and it can be difficult to convince the client of the value of commissioning. The financial benefits are well documented in *The Cost-Effectiveness of Commercial-Buildings Commissioning*, a cost-effectiveness study of over 200 commercial buildings, by the Lawrence Berkeley National Laboratory. They found the savings often paid for the commissioning in the first year. The web site with access to the complete text is given towards the end of this chapter, in "Resources".

Occupant satisfaction is often dependent on how quickly a problem is resolved, which brings us to the issue of warranties. How quickly must the contractor respond to a problem during working hours and during off hours? What personnel and replacement parts must be available locally or in a stated time? This will vary depending on the project size and location. For remote locations, having dial up or internet access to the system can be very useful in helping the contractor send the correct spare parts with the service personnel.

Manufacturers update their software to deal with errors and also to make significant performance changes. Some designers require that the installer installs any software updates as they become available during the warranty period. This is good in that the client has the latest version, but it can have some unexpected consequences. On one project the load shedding software was changed, so the update did not run the load shedding correctly. The installer and client did not realize the change until the next electrical bill was more than double what it would have been if the load shedding had continued. In another case, the historical data collection software was changed and the history became inaccessible. Be careful, a good working system is not necessarily better having updated software installed.

Warranty periods are typically one year from demonstration that the system is in complete working order. This is even if the occupants have been receiving partial, or full, use from the system. Most electronic failures occur in the first year and an extended warranty is generally not economic as a warranty for component failure.

Many specifications include a provision that the manufacturer continue to have replacement components available for five years from completion. This warranty requirement does not necessarily include the replacement, only that they are available. Since there is no financial provision in the contract for this requirement, it is difficult to enforce unless the contract is covered by a performance bond. A performance bond is issued by an insurance company, typically, to guarantee the completion of a project but it may also be required to cover the extended warranty period.

12.7 Resources

The most valuable resource is the manufacturers. They know their products and want them used successfully. Ask questions and go on asking until you understand. Manufacturers have their own names for things that you may not recognize: ask them to explain. Remember that a contract that runs smoothly without any nasty surprises is best for everyone. Many organizations and manufacturers' web sites have very good technical literature available to view and download.

The following are some specific web site suggestions to help you get started. They are no particular order.

BACnet International is the association of those interested in the use and promotion of BACnet: http://www.bacnetassociation.org

CtrlSpecBuilder.com is a free site for generating DDC specifications and drawings for common HVAC systems: http://www.ctrlspecbuilder.com/

Automated Buildings' web site has both information and many useful links: http://www.automatedbuildings.com

LonMark International is the organization devoted to the use and development of LonWorks. Both general introductory and detailed technical information is available on their site: http://www.lonmark.org

The Continental Automated Buildings Association focuses on building automation with significant focus on residences: http://www.caba.org

EPA Portfolio Manager is an USA program designed to provide comparative energy use information about most types of commercial buildings and information, available at: http://www.energystar.gov/index.cfm?c=evaluate_ performance.bus_portfoliomanager

Whole Building Diagnostician (WBD) software field test evaluation, available at http://www.buildingsystemsprogram.pnl.gov/fdd/publications/ PNNL-39693.pdf

The Cost-Effectiveness of Commercial-Buildings Commissioning, Evan Mills, Lawrence Berkeley National Laboratory, 2004, is a report on a large investigation into the value of commissioning commercial buildings. The full text is available at: http://eetd.lbl.gov/emills/PUBS/Cx-Costs-Benefits.html

The following are specifically relevant, for purchase, documents that are mentioned in the text:

National Electrical Code NFPA 70: 2008 is the latest rules for electrical installation in residential, commercial, and industrial occupancies in the USA: http://www.nfpa.org

ASHRAE Guideline 0 – *The Commissioning Process* 2005. The purpose of this guideline is to describe the Commissioning Process capable of verifying that the facility and its systems meet the owner's project requirements. The procedures, methods, and documentation requirements in this guideline describe each phase of the project delivery and the associated commissioning processes from pre-design through occupancy and operation, without regard to specific elements, assemblies, or systems: http:/www.ASHRAE.org

ASHRAE Guideline 1- The HVAC Commissioning Process 1996 – The procedures, methods, and documentation requirements in this guideline cover each phase of the commissioning process for all types and sizes of HVAC systems, from pre-design through final acceptance and post-occupancy, including changes in building and occupancy requirements after initial occupancy. The guideline provides procedures for the preparation of documentation of the owner's assumptions and requirements; design intent, basis of design, and expected performance; verification and functional performance testing; and operation and maintenance criteria. This guideline includes a program for training of operation and maintenance personnel: http:/www.ASHRAE.org

ASHRAE GPC-13-2007 *Guideline to Specifying DDC Systems.* The purpose of this guideline is to provide recommendations for developing specifications for direct digital control (DDC) systems in heating, ventilating, and air-conditioning (HVAC) control applications. The guideline includes discussion of options and sample specification language: www.ashrae.org (product code 86823).

ASHRAE Guideline 14-2002 *Energy Management, Energy Savings.* This document provides guidelines for reliably measuring energy and demand savings of commercial equipment. Energy service companies (ESCOs), ESCO customers, utilities and others can use these measurements before the sale or lease of energy-efficient equipment to determine post-transaction savings. These measurements can also be used to document energy savings for various credit programs, such as emission reduction credits associated with energy efficiency activities.

ASHRAE Guideline 22-2008 The purpose of this guideline is to enable the user to continuously monitor chilled-water plant efficiency in order to aid in the operation and improvement of that particular chilled-water plant. The focus is on the various available levels of instrumentation and the collection of data needed for monitoring the efficiency of an electric-motor-driven central chilled-water plant.

ASHRAE Standard 105-2007 *Standard Methods of Measuring, Expressing and Comparing Building Energy Performance.* This standard is intended to foster a commonality in reporting the energy performance of existing or proposed buildings to facilitate comparison, design and operation improvements, and development of building energy performance standards. It provides a consistent method of measuring, expressing, and comparing the energy performance of buildings. This latest revision provides a method of energy performance comparison that can be used for any building, proposed or existing, and that allows different methods of energy analysis to be compared.

ASHRAE – Standard 135-2004, *BACnet® – A Data Communication Protocol for Building Automation and Control Networks Systems* (ANSI approved), available from ASHRAE Bookstore at http://www.ashrae.org bookstore (product code 86449). It is also ISO 16484-5:2003 Building automation and control systems – Part 5: Data communication protocol, available from the International Organization for Standardization, Geneva 20, Switzerland: http://www.iso.org

The purpose of this standard is to define data communication services and protocols for computer equipment used for monitoring and control of HVAC&R and other building systems, and to define an abstract, object-oriented representation of information communicated between such equipment, thereby facilitating the application and use of digital control technology in buildings. Note that this is a very, very technical standard and not easy reading, as it is dealing with computer programming. To get started with understanding BACnet in any detail go to the site for the committee that produces the BACnet Standard: http://www.bacnet.org

Bibliography

Alerton Controls (2004) *Use of Controller in DPAC Application.* Redmond, WA, www. alerton.com.

ASHRAE Guideline 13-2000 – Specifying Direct Digital Control Systems, ASHRAE, Atlanta, GA. June 2000 + Addenda 2005.

ASHRAE Guideline 22-2008 Instrumentation for Monitoring Central Chilled Water Plant Efficiency. ASHRAE Atlanta, GA.

ASHRAE Standard 180 (2006) *Standard Practice for Inspection and Maintenance of HVAC Systems.* Atlanta, GA.

Bailey, H. (2005) *Fungal Contamination: A Manual for Investigation, Remediation, and Control.* Jupiter, FL: BECi.

BBJ Environmental Solutions, Baker, R. (2005) "Coil cleaning and its resultant eenergy savings and maintenance enhancements," *FWC ASHRAE* newsletter, Tampa, November.

Behnken, J. (2001) "Building Automation for Multisite Facilities," *ASHRAE Journal,* June, p. 53.

Brambley, M., Chassin, D., Gowri, K., Kammers, B., Branson, D., (2000) "DDC and the WEB," *ASHRAE Journal,* December, p. 38.

Chapman, R. (2001) "How interoperability saves money," *ASHRAE Journal,* February, p. 44.

Fisher, D. (2000) "BACnet at work," *ASHRAE Journal,* July, p. 34.

Hahn, W. and Michael, P.E. (1999) "Fast track HVAC conversion," *ASHRAE Journal,* December, p. 44.

Honeywell HB&C, Automation and Control Specialist and LCBS CD-Rom. Minneapolis, MN.

Hegberg, M. (1996) "MCH4 Controllers – HVAC-1 Class Presentation," CD-Rom.

Hydeman, M., Seidl, R. and Shalley, C. (2005) "Data center commissioning," *ASHRAE Journal,* April, p. 60.

Invensys Inc. (2002–4) *DDC Systems Information.* Rockford, IL.

Johnson Controls Inc. (2002–4) Control Systems Information. Milwaukee, WS.

Otto, K. (2001) "Maximizing the value of open systems," *ASHRAE Journal,* May, p. 33.

Persily, A. (2002) "New Requirements of Standard 62-2001," *ASHRAE IAQ Applications,* Fall, p. 8.

Seem, J., House, J. and Monroe, R. (1999) "On-line monitoring and fault detection," *ASHRAE Journal,* July, p. 21.

Schulte, R. Bridges, B. and Grimsrud, D. (2005) "Continuous IAQ monitoring," *ASHRAE Journal,* May, p. 38.

Shadpour, F. and Willis, L.P.E. (2001) "DDC systems as a tool for commissioning," *ASHRAE Journal,* November, p. 43.

Wilkinson, R.J. (2002) *Commissioning and Controls.* HPAC Engineering: Cleveland, OH.

Wiggins, S.R. (2005) "Retrocommissoning," *ASHRAE Journal,* March. p. 76.

Index

A

Across-the-line magnetic motor
    starter, 46–47
Adjustable speed drives (ASDs).
    See Variable speed drives (VSDs)
Air-conditioning plant, 117–118, 217
Air dryers, 197
Airflow-measuring device, 238
Air flow measuring station (FMS), 146
Airfoil dampers, 87–88
Air handling unit (AHU), 226–228,
    237–238, 266–267
Air-heating system, 5–6
Air temperature, 234
Air valve, 164–165
Alternating current (AC) circuits
  capacitor, DC circuit, 36
  circuit diagram, 35–36
  definition, 35
  impedance, 38
  inductor, DC circuit, 37–38
  magnetic field, 36–37
  power equation, three-phase
    circuit, 39
  power factor, 38–39
Analog control. *See* Modulating control
Analog electronic controls
  actuator, 211–212
  auxiliary devices, 212
  economizer enthalpy control, 210
  idealized operational amplifier
    (Op-amp), 206–207
  integral and derivative Op-amp, 208
  proportional Op-amp, 207
  sensor, 206
  single-zone unit controllers, 210

  summing and subtraction
    Op-amp, 208–209
  VAV box, 211
Analog to digital (A/D)
    converter, 255–256
AND-gate, 266
Annubar® flow sensors, 139–140
ANSI/ASHRAE Standard 135-2004, 302
Application software, 260–261, 268
Application-specific programming, 261,
    268
ARCNET standards, 299
Artificial intelligence, 23
ASHRAE Standard 62.1, 241–242,
    245–246
Aspirated (fan powered)
    psychrometers, 132–133
Authority distortion, 83–84
Automatic control, 2–3
Auxiliary devices, 178–179, 212
  CO_2 and indoor air quality
    (IAQ), 156–157
  discriminator relay, 191–192
  electric and pneumatic transducer, 158
  electric/pneumatic switch
    application, 194–195
  fan and pump status switches, 149–151
  fire and smoke detectors, 154–155
  fluid level device, 158–159
  indicating light and audible alarm, 155
  IP transducer–bleed type, 195
  isolation room pressure, 156
  limit and manual switches, 152
  low/high limit switches, 153–154
  manual pneumatic switch, 195–196
  manual rotary switch, 152–153